Classical Mechanics

J. W. LEECH, BSc, PhD

Chairman, Department of Physics,
University of Waterloo, Ontario

LONDON NEW YORK

CHAPMAN AND HALL

First published 1958 by Methuen & Co Ltd
Reprinted once
Second edition 1965
First published as a Science Paperback 1965
by Chapman and Hall Ltd
11 New Fetter Lane, London EC4P 4EE
Published in the USA by Chapman and Hall
29 West 35th Street, New York NY 10001
Reprinted 1968, 1970, 1978, 1982, 1987

Printed in Great Britain by
J. W. Arrowsmith Ltd, Bristol

ISBN 0 412 20070 8

To W. E. K.
for his initial stimulation
of interest

Preface

It is a feature of the history of the subject that the study of atomic physics was accompanied by a partial neglect of that of classical mechanics. This led to the unsatisfactory situation in which the physicist was expected to assimilate the elements of quantum and statistical mechanics without understanding the classical foundations on which these subjects were built. The situation has improved in recent years through the general lengthening of degree courses, and it is now usual to study the analytical formulation at the late under-graduate stage. A number of excellent treatises are available, and there are also many elementary accounts to be found in general works on physical principles. However, there has been available so far no self-contained introduction to the subject which provides the beginner with a broad general review without involving him in too much detail. It is hoped that this book may bridge the gap by pro-viding the experimental physicist with a sufficient background for his theoretical understanding and the theorist with some stimulus to study the masterpieces of the subject.

The mathematical equipment required is no more than in the normal honours physics course. For the purposes of Chapters IX–XI it includes an elementary knowledge of cartesian tensors. A familiarity with Newtonian mechanics and some acquaintance with special relativity theory are presumed, though summarizing accounts are also given. An indication has been made of the method of tran-sition to quantum mechanics, since, without this, some of the classical development seems merely an elegant but pointless in-tellectual exercise. No mention has been made of statistical me-chanics as the ground work for this has obvious applications in other ways. It was thought highly desirable to include an account of the analytical formalism as applied to continuous systems with special reference to fields. The study of the latter is an immensely important part of modern physics, but connected accounts of the classical foundations are not easily accessible.

My warmest thanks are due to Dr R. O. Davies, Dr A. J. Manuel and Mr K. A. Woolner for their tireless criticisms and encouragement in the course of the preparation of the manuscript.

J. W. L.

Queen Mary College
March 1958

Preface to the Second Edition

The appearance of a second edition of this book does not seem to call for any revision of its structure. It is of interest, however, to note two recent shifts of emphasis in the applications of analytical mechanics. These are the increasing use of quantum field methods in solid state physics on the one hand and the development of supplementary approaches to problems in elementary particle physics.

Advantage has been taken of the new printing to incorporate a collection of examples with which the reader can test his understanding. For these I am chiefly indebted to my erstwhile colleague Professor J. G. Powles. A few new references have been added and others brought up to date; the text is otherwise unaltered apart from the correction of some minor errors.

J. W. L.

Queen Mary College
April 1965

Contents

CHAPTER I

Introduction

Classical mechanics concerns itself initially with describing the motions of entities known as *particles*. A complete description of a particle at any instant of time is obtained by specifying three space co-ordinates together with a scalar constant termed its mass. Although it is impossible rigorously to identify any real subdivision of matter with such a concept, the motions of bodies of macroscopic size can be described very accurately by regarding them as collections of particles in the above sense.

Particles are held to move in accordance with Newton's laws which may be stated as follows:

(1) *A body continues in a state of rest or moves with uniform speed along a straight line unless subjected to an external force.*

(2) *The rate of change of momentum of a body is proportional to the magnitude of the external force acting upon it and takes place in the direction of that force.*

(3) *The forces exerted by two bodies on each other are equal in magnitude and opposite in direction.*

A number of subsidiary definitions are required of the concepts used. The final scheme may readily be criticized on the grounds of circularity since it may be argued that the postulates involve the terms *mass, force, straight line*, etc., which have no satisfactory independent definition. Bridgman, following Mach, has proposed operational definitions of mass and force which seem to overcome part of this difficulty though the problem of identifying a straight line is left untouched. An important point of principle is involved, but it is not proposed to discuss it further here.* Instead, it will be assumed that

* For a thorough discussion of the problem see the works by Mach, Bridgman and Lindsay and Margenau quoted in the bibliography at the end of the book.

the terms in question have some intuitively understood meaning and that Newton's laws are logical statements concerning their inter-relations.

The aim will be to show that Newton's laws can be replaced by a single postulate (the *variational principle*) which is, for many purposes, more convenient to handle. In the realms of particle mechanics this alternative postulate is completely equivalent to the assumption of Newton's laws. Its virtue lies in the ease with which it can be used to formulate complex problems.

The scheme based on Newton's laws can be termed *vectorial mechanics* since it deals with quantities such as force, velocity, etc., which are essentially vectorial in character. The alternative scheme, initiated by Leibnitz and associated with the names of Euler, Lagrange and Hamilton, is referred to as *analytical mechanics*. The fundamental quantities are now scalars rather than vectors and the dynamical relations are obtained by a systematic process of differentiation.

In the simplest examples the use of the analytical method may be compared to chartering an aircraft to cross the road. In such cases it must be realized that the purpose is to gain familiarity with the new approach. Once this familiarity has been established the way is open to an appreciation of the greater convenience of the analytical method in the formulation of more complicated problems. A further point is that the new method is capable of generalization to situations such as classical field theory and quantum mechanics in which Newton's laws are inapplicable.

From an aesthetic point of view a case could be made out for first postulating the laws of mechanics in their most general (analytical) form and then showing how Newton's laws emerge under certain restrictive conditions. This programme is acceptable once the alter-native formalism has been assimilated, but it is not the best approach when there is an existing familiarity only with Newton's laws.

It must be emphasized that, where both apply, the difference be-tween the vectorial and the analytical method lies in the process of *formulating* the equations of motion. These are normally a set of differential equations, and in each case the final stage of solution requires a knowledge of the techniques of handling such equations.

On this account the examples quoted will normally be developed only to the stage at which the equations of motion appear.

A warning of this sort is necessary in order to avoid the type of criticism which is sometimes levelled at the study of vector analysis. Too often the erroneous impression is gained that vector methods can be used to solve all problems in detail; the subsequent disillusionment can lead to the feeling that their importance has been grossly exaggerated. The truth is, of course, that the use of vectors usually results in a considerable saving of mental effort in the translation of the physical circumstances of a problem into mathematical terms. In all but a few elementary cases it is necessary at some stage to resolve the vectors into their components; the art of the matter then consists in selecting the most useful type of co-ordinate system. A similar situation exists in analytical mechanics.

CHAPTER II

Fundamental Ideas

This chapter is intended as a brief summary of those aspects of mechanics which stem immediately from Newton's laws and which are particularly important in the development of the Lagrangian and Hamiltonian formulations.

According to Newton's laws the motion of a *system* of N particles is represented by the solution of a set of N vector equations of motion of the type:

$$\mathbf{F}_s = \frac{d}{dt}\mathbf{p}_s = \frac{d}{dt}(m_s\dot{\mathbf{r}}_s) \qquad (s = 1, 2, 3, \ldots N) \tag{2.1}$$

These may be written alternatively as a set of $3N$ scalar equations of the form:

$$F_i = \frac{d}{dt}p_i = \frac{d}{dt}(m_i\dot{x}_i) \qquad (i = 1, 2, 3, \ldots 3N) \tag{2.1'}$$

where $\dot{x}_i \equiv \dfrac{dx_i}{dt}$, and it is assumed that the position of each particle is fixed by three cartesian type co-ordinates x_i. The labelling is such that the three co-ordinates of any particle have suffices which run serially. The first particle has co-ordinates x_1, x_2 and x_3, and its mass is correspondingly $m_1 = m_2 = m_3$. Initially only cartesian co-ordinate systems will be considered.

To determine the motion of the system it is necessary to specify the force components and the *initial values* or *boundary conditions*. In principle, the forces could depend in any way on the time, the co-ordinates and their derivatives of any order. However, it is a matter of experimental observation that, where the forces can be expressed in analytical form, they depend at most only on the position co-ordinates and the velocities. The equations of motion are then second-order differential equations and require for their solution the specification of two boundary conditions for each variable x_i. The conditions are normally given as values of the x_i and of the $\dot{x}_i$ at some particular instant; they may then be specified in an arbitrary manner. More rarely they may take other forms, such as the values of the x_i at each of two instants t_1 and t_2; in this case incompatibility with the given force components may prevent the boundary conditions being assigned in a completely arbitrary manner. In all cases a total of $6N$ constants are required to be specified.

Conservation Laws

It is often convenient to recognize that the force acting on each particle of a system may be regarded as the sum of two parts:

$$\mathbf{F}_s = \mathbf{F}_s^{(i)} + \mathbf{F}_s^{(e)} \tag{2.2}$$

where $\mathbf{F}_s^{(i)}$ is that part due to the other particles of the system and $\mathbf{F}_s^{(e)}$ originates outside the system. The point of the distinction is that, if the system is considered as a whole, Newton's third law may be invoked to show that the internal forces cancel in pairs, i.e.:

$$\mathbf{F} \equiv \sum_{s=1}^{N} \mathbf{F}_s = \sum_{s=1}^{N} \mathbf{F}_s^{(e)} \tag{2.3}$$

Combining equations (2.1) and (2.3) gives:

$$\sum_{s=1}^{N} \mathbf{F}_s^{(e)} = \sum_s \frac{d}{dt}(\mathbf{p}_s) = \frac{d}{dt}\left\{\sum_s \mathbf{p}_s\right\} = \frac{d}{dt}\,\mathbf{P} \tag{2.4}$$

where $\mathbf{P}$ is the total momentum of the system.

If the sum of the external forces acting on the system is zero it follows that $\mathbf{P} =$ constant; i.e., *that the total linear momentum of the system is conserved*.

Another quantity of interest is the *angular momentum*. For the particle s the angular momentum about the origin of co-ordinates is defined as:

$$\mathbf{M}_s = \mathbf{r}_s \wedge \mathbf{p}_s = \mathbf{r}_s \wedge m_s \dot{\mathbf{r}}_s \tag{2.5}$$

From (2.1):

$$\mathbf{r}_s \wedge \mathbf{F}_s = \mathbf{r}_s \wedge \frac{d}{dt}(m_s\dot{\mathbf{r}}_s) = \frac{d}{dt}(\mathbf{r}_s \wedge m_s\dot{\mathbf{r}}_s) = \frac{d}{dt}\mathbf{M}_s \qquad (2.6)$$

giving the result that the rate of change of the angular momentum of the particle s is equal to the moment of the force applied to it.

Summing over all the particles of the system:

$$\sum_s \mathbf{r}_s \wedge \mathbf{F}_s = \sum_s \frac{d}{dt}\mathbf{M}_s = \frac{d}{dt}\sum_s \mathbf{M}_s = \frac{d}{dt}\mathbf{M} \qquad (2.7)$$

The rate of change of the total angular momentum is thus equal to the sum of the moments of all the forces, both internal and external, acting on the system. The cancellation of the effects due to the internal forces no longer occurs simply on the basis of Newton's third law. It requires instead the more restrictive condition that they be *central forces*, i.e., that they are directed along the lines joining the particles. With this proviso (2.7) reduces to:

$$\sum_s \mathbf{r}_s \wedge \mathbf{F}_s{}^{(e)} = \frac{d}{dt}\mathbf{M} \qquad (2.7')$$

If, further, the total moment of the external forces is zero it follows that $\mathbf{M} = \text{const.}$, i.e., that the *total angular momentum of the system is conserved*.

The above conservation theorems play an important role in mechanics. In some cases the identification of conserved quantities may be regarded as the solution to a problem.

Energy

In seeking for a measure of the total effect of a force it seems natural to consider its line integral taken over the path of the particle upon which it acts. From (2.1) we have:

$$\int_1^2 \mathbf{F}_s \cdot d\mathbf{r}_s = \int_1^2 \frac{d}{dt}(m_s\dot{\mathbf{r}}_s) \cdot d\mathbf{r}_s = \left[\tfrac{1}{2}m_s\dot{\mathbf{r}}_s{}^2\right]_1^2 \qquad (2.8)$$

This line integral is a scalar quantity and is referred to as the *work done by the force*. In line with a general desire to consider effects in terms of conserved quantities it is assumed that this work is stored up in the particle as the *kinetic energy of motion*. The quantity $\frac{1}{2}m_s\dot{\mathbf{r}}_s^2$ appearing on the right-hand side of (2.8) is thus defined as the kinetic energy T_s of the particle s.

Summing over all the particles of the system we now have:

$$\int_1^2 \sum_{s=1}^{N} \mathbf{F}_s . d\mathbf{r}_s = \sum_s \int_1^2 \mathbf{F}_s . d\mathbf{r}_s = \sum_s \left[T_s\right]_1^2 = T^{(2)} - T^{(1)} \quad (2.9)$$

where $T^{(2)}$ and $T^{(1)}$ denote the final and initial values of the total kinetic energy of the system.

In certain cases $\displaystyle\sum_{s=1}^{N} \mathbf{F}_s . d\mathbf{r}_s$ may be expressed as a perfect differential $- dV$, where $V = V(x_i)$ is a function of the position co-ordinates x_i. The work done by the forces is then independent of the actual paths followed by the particles, and depends only upon their initial and final positions. Such systems are said to be *conservative*. With this restriction (2.9) gives:

$$V^{(1)} + T^{(1)} = V^{(2)} + T^{(2)} \quad (2.10)$$

This relation is interpreted as stating that the total energy, $E = T + V$, of the system is conserved; though it may be exchanged between the kinetic and potential forms T and V.

It can be argued that, whereas there appears to be some 'reality' to kinetic energy, the same cannot be said of potential energy. In a sense the latter is a fictitious quantity so defined that changes in its value exactly compensate any changes in the kinetic energy. The fact that the sum of the two quantities is conserved is then no cause for surprise. However, similar objections can be raised to the 'reality' of forces since these can also be regarded merely as balancing terms in mathematical equations. The resolution of the difficulty is in both cases through the realization that the ultimate object in mechanics is to describe the motion of a system. If it assists in any way in achieving this aim, the relabelling of any well defined quantities associated with the motion is a permissible act. In the case of the various quantities labelled energy there can be no doubt that this does materially assist the process of description.

The practical importance of the energy concept lies in the fact that all the mechanical properties of complex systems may be summarized by specifying the mathematical form of a limited number of scalar energy functions. Analytical mechanics consists of a general development of this idea.

The 'Principle' of Minimum Energy

In the previous section a conservative system was defined effectively as one in which the force components are given by:

$$F_i = -\frac{\partial V}{\partial x_i} \quad \text{where } V = V(x_i) \tag{2.11}$$

If the system is in equilibrium:

$$F_i = 0 \quad \text{i.e., } \frac{\partial V}{\partial x_i} = 0 \quad (i = 1, 2, 3, \ldots 3N) \tag{2.12}$$

However, these $3N$ conditions are the requirements that the function V shall have a stationary value. In equilibrium, therefore, the potential energy of a conservative system has a stationary value. The term *minimum energy principle* follows from the fact that the equilibrium considered is normally a stable one, but there is no general necessity for this.

The result is usually referred to as a 'principle' but, as has been seen, it is an immediate consequence of the definition of a conservative system. It constitutes an elementary form of the method of analytical mechanics, applicable only to the rather restricted range of problems concerning static equilibrium in conservative systems.

Constraints

It is possible that the motion of a system may be limited in some way by *constraints*. These may be defined as geometrical restrictions on the motion. An elementary example is that of a simple pendulum in which the motion is limited to the extent that the pendulum bob (idealized as a particle) remains at a constant distance l from a fixed point **a** (the point of suspension). This is an example of a *holonomic constraint* which can be represented by its *equation of constraint*:

$$(\mathbf{r} - \mathbf{a})^2 = l^2$$

where **r** is the position vector of the bob. The essential feature of a holonomic constraint is the existence of one or more *algebraic* relations connecting some or all of the space and time co-ordinates of the particles of the system. Other examples are a particle moving over a plane (not necessarily fixed in space) and an Atwood's machine in which two particles are connected by a string of constant length. A rigid body is a particularly simple type of holonomically constrained system in which the separations of all the particles are maintained constant.

Not all constraints are holonomic. A particle could, for instance, be placed on the surface of a solid sphere, its subsequent motion would then be governed by the inequality $(\mathbf{r} - \mathbf{r}_0)^2 > a^2$ expressing the fact that its distance from the centre (whose position vector is $\mathbf{r}_0$) cannot be less than the radius of the sphere. A disk rolling on a horizontal plane without slipping and with its axis horizontal is also a non-holonomic system since the constraint is represented by a non-integrable differential relation involving the co-ordinates of its point of contact with the plane. (This example is considered further in Chapter VI.) Other types of non-holonomic constraint may be quite impossible to represent in mathematical terms. Standard procedures for the determination of the motion of a non-holonomic system can be devised only when there are differential equations of constraint. Other cases require individual treatment and may not always be soluble.

The effect on a system of having m equations of constraint is that m of the original $3N$ variables describing the system become dependent, rather than independent, variables. It is then said to possess only $(3N - m)$ *degrees of freedom*. Difficulties sometimes arise in that it may not be obvious which are to be regarded as the dependent variables; as will be seen later, these may be overcome by a symmetrical treatment involving undetermined multipliers.

In general, forces are required to maintain the constraints in a system. These forces are not known initially. A knowledge of their values is not often important and they are normally eliminated from the equations of motion at an early stage, but methods can be devised for their determination if required (see Chapter VI). If the conditions of constraint are independent of time the associated forces of constraint, although not necessarily constant, do no work in the course of the motion. On this account such constraints are usually termed *workless*. Time-dependent constraints are not workless.

Constraints are idealized concepts introduced to assist in the solution of mechanical problems. Generally they represent highly simplified forms of complex systems. This is the case with the problem of the particle moving over a horizontal plane. In a real case the plane would be the surface of an elastic solid and would be slightly deformed under the weight of the object. In the process of idealization to a geometrical plane the continuous real forces of reaction are replaced by discontinuous constraint forces. Such discontinuous forces are not envisaged in Newton's laws and necessitate a new postulate (represented by eqn. 2.18) before they can be included in a convenient mathematical scheme.

Generalized Co-ordinates

In the great majority of mechanical problems little progress towards a solution can be made without resort to a particular co-ordinate system. In view of their simplicity, cartesian systems are used wherever possible, but they are not always suitable. In considering the motion of a particle under a central force, for instance, it is much more useful to employ a plane polar co-ordinate system.

Systems alternative to cartesian co-ordinates will be considered in a general manner so that they may later be particularized in any convenient way. The variables will normally be distances or angles, but others are possible—particularly in later extensions of the formalism of classical mechanics. Expressed in terms of *generalized co-ordinates* the equations of motion have the same general appearance, but involve terms about which there may be some debate as to whether they may rightfully be classified as 'force' terms or 'rate of change of momentum' terms. Examples of these are centrifugal and Coriolis 'forces', both of which are associated with a rotating co-ordinate system. Neither of these is associated with any external effect; they are fictitious forces arising from the method of description as peculiarities of the co-ordinate system employed. These *fictitious force* terms complicate the formulation of the equations of motion very considerably in the vectorial approach. It is one of the great advantages of the analytical method that they appear quite automatically as the product of a systematic mathematical process.

An advantage of generalized co-ordinates is that constraints may often be automatically allowed for by a suitable choice of system. This eliminates the need for writing out separately the $3N$ equations of motion and the m equations of constraint. This consideration is well exemplified in the case of a rigid body, whose motion may be described in terms of three co-ordinates (possibly still cartesian) representing the position of the centre of mass together with the specification of three angles.

Configuration Space

The determination of the motion of a single particle constitutes a mechanical problem in three dimensions. Two particles can be considered each to be described by a set of three co-ordinates; alternatively, a problem involving two particles can be regarded as one of a single particle moving in a six-dimensional space. The difference is, in a sense, one of terminology only and philosophical difficulties are readily removed by referring to six variables instead of six dimensions. Nevertheless, the geometrical description can be very illuminating and helpful since quite simple diagrams may be used to illustrate the argument.

In general, a problem involving N particles can be treated as one of a single particle moving along a *trajectory* in $3N$-dimensional space. This space is referred to as *configuration space*. Constraints are said to restrict the motion to a *sub-space* of appropriately fewer dimensions. The terminology may be used in connection with any type of generalized co-ordinate system.

Virtual Work and d'Alembert's Principle

It is possible and often desirable to imagine arbitrary instantaneous alterations in the position vectors of the particles of a system. The purpose of such considerations, known as *virtual displacements*, is to provide a kind of mathematical probe in order to investigate the properties of the system. An infinitesimal virtual displacement of the representative sth particle will be denoted by $\delta \mathbf{r}_s$.

If the system is in equilibrium then, by definition:

$$\mathbf{F}_s = 0 \qquad (s = 1, 2, \ldots N) \qquad (2.13)$$

from this it follows that:

$$\mathbf{F}_s . \delta \mathbf{r}_s = 0 \qquad (2.14)$$

and also that:

$$\sum_s \mathbf{F}_s . \delta \mathbf{r}_s = 0 \qquad (2.15)$$

Results (2.14) and (2.15) are trivial mathematical deductions from the definition of equilibrium. If, however, the forces $\mathbf{F}_s$ are continuous functions of position, then (2.15) takes on a new physical meaning which can be expressed symbolically as:

$$\delta W = 0 \qquad (2.16)$$

This result represents a statement of the *principle of virtual work* which asserts:

'*The work done is zero in the course of an arbitrary infinitesimal virtual displacement of a system from a position of equilibrium.*'

If constraints are present in the system, then the forces may be classified either as *applied* (or *impressed*) *forces* ($\mathbf{F}_s^{(a)}$) or as *forces of constraint* ($\mathbf{F}_s^{(c)}$), thus:

$$\mathbf{F}_s = \mathbf{F}_s^{(c)} + \mathbf{F}_s^{(a)} \qquad (2.17)$$

(2.15) then becomes:

$$\sum_s \mathbf{F}_s^{(c)} . \delta \mathbf{r}_s + \sum_s \mathbf{F}_s^{(a)} . \delta \mathbf{r}_s' = 0 \qquad (2.15')$$

It is not possible to interpret (2.15') as a statement concerning virtual work since the discontinuous nature of constraint forces is inherent in the concept of constraints. The difficulty is surmounted by the introduction of a *new postulate*:

$$\mathbf{F}_s^{(c)} . \delta \mathbf{r}_s > 0 \qquad (2.18)$$

for all $\delta \mathbf{r}_s$ which are compatible with the constraints.

This is really a statement concerning the *directions* of possible virtual displacements relative to the forces of constraint. Its plausibility may be tested by considering elementary examples of constrained systems such as that of a particle at rest on a horizontal plane; in this case the only displacements of the particle consistent with the constraint are along the plane or in an upwards direction.

Combining (2.15′) and (2.18) yields:

$$\sum_s \mathbf{F}_s^{(a)} . \delta \mathbf{r}_s < 0 \qquad (2.19)$$

The only forces now involved in the statement are the applied ones, which may be assumed continuous in space. It is once more possible, therefore, to interpret the result as a statement concerning work done by these applied forces in the course of a displacement consistent with the constraints:

$$\delta W \equiv \sum_s \mathbf{F}_s^{(a)} . \delta \mathbf{r}_s < 0 \qquad (2.19')$$

If, now, the virtual displacements considered are confined to those which are, in the geometrical sense, reversible (denoted by the symbol δ'), then (2.19) implies:

$$\sum_s \mathbf{F}_s^{(a)} . \delta' \mathbf{r}_s < 0$$

and

$$\sum_s \mathbf{F}_s^{(a)} . (- \delta' \mathbf{r}_s) < 0$$

These statements are only compatible if:

$$\sum_s \mathbf{F}_s^{(a)} . \delta' \mathbf{r}_s = 0 \qquad (2.20)$$

This is a *generalized form of the principle of virtual work* which states:

'*The work done is zero in the course of an infinitesimal reversible virtual displacement, compatible with the constraints, of a system from a position of equilibrium.*'

Since there are constraints the $\delta' \mathbf{r}_s$ will not now be all independent. It follows that, unlike the position with respect to (2.15), it is not possible to deduce from (2.20) that $\mathbf{F}_s^{(a)} = 0$. In many cases, however, it is possible to transform to a generalized co-ordinate system (denoted by q_i) in which the constraint conditions are automatically allowed for. The relation (2.20) can then be reduced to:

$$\sum_i Q_i^{(a)} \delta' q_i = 0 \qquad (2.20')$$

In this case the number of generalized co-ordinates is identical with the number of degrees of freedom of the system, i.e., the generalized virtual displacements $\delta' q_i$ may all be made independently. It is then possible to deduce the fact that:

$$Q_i^{(a)} = 0 \qquad (2.21)$$

Considerations so far have covered only systems in static equilibrium. Systems in motion may be included by an extension of the argument. The equations of motion were earlier stated to be:

$$\mathbf{F}_s = \frac{d}{dt}(m_s \dot{\mathbf{r}}_s) \qquad (2.1)$$

i.e.,

$$\mathbf{F}_s - \frac{d}{dt}(m_s \dot{\mathbf{r}}_s) = 0$$

For an arbitrary virtual displacement it follows, as before, that:

$$\sum_s \left(\mathbf{F}_s - \frac{d}{dt}(m_s \dot{\mathbf{r}}_s) \right) . \delta \mathbf{r}_s = 0 \qquad (2.22)$$

The $\mathbf{F}_s$ may be divided into constraint and impressed forces giving:

$$\sum_s \mathbf{F}_s^{(c)} . \delta \mathbf{r}_s + \sum_s \left(\mathbf{F}_s^{(a)} - \frac{d}{dt}(m_s \dot{\mathbf{r}}_s) \right) . \delta \mathbf{r}_s = 0 \qquad (2.22')$$

and, with the displacements restricted to be compatible with the constraints, it can again be *postulated* that:

$$\mathbf{F}_s^{(c)} . \delta \mathbf{r}_s > 0 \qquad (2.18)$$

whence:

$$\sum_s \left(\mathbf{F}_s^{(a)} - \frac{d}{dt}(m_s \dot{\mathbf{r}}_s) \right) . \delta \mathbf{r}_s < 0 \qquad (2.23)$$

The further restriction to reversible virtual displacements then gives:

$$\sum_s \left(\mathbf{F}_s^{(a)} - \frac{d}{dt}(m_s \dot{\mathbf{r}}_s) \right) . \delta' \mathbf{r}_s = 0 \qquad (2.24)$$

Again, it is not possible immediately to deduce anything concerning the bracketed quantities, since the presence of the constraints prevents the displacements being made independently. This difficulty may again be surmounted in some cases by transforming to a suitable generalized co-ordinate system. The general method of doing so will be considered in the next chapter as an essential part of the analytical method.

The first term of (2.24) is again identifiable as work done in the course of the virtual displacement. The status of the second term is rather more debatable. It is usual to rewrite $- \dfrac{d}{dt}(m_s \dot{r}_s)$ as I_s and to relabel it a *force of inertia*. (2.24) then becomes:

$$\sum_s (\mathbf{F}_s^{(a)} + \mathbf{I}_s) \cdot \delta' \mathbf{r}_s = 0 \qquad (2.25)$$

The combined applied and inertial forces are known as the *effective* forces. With these new definitions it may be asserted that:

'*The total work done by the effective forces is zero in a reversible infinitesimal virtual displacement, compatible with the constraints, of any dynamical system.*'

This is known as *d'Alembert's principle*. Its verbal formulation was prompted by the somewhat artificial device of re-defining certain terms as forces of inertia and it is doubtful whether the expression 'work done' here has any real significance. This is, however, an unimportant quibble. The assertion of the principle implies the truth of the relation (2.24). The latter is probably the most general summarizing statement in the whole of the mechanics of material systems and all later statements of principle, including Hamilton's, are derivable from it.

The Newtonian equations of motion are also deducible from (2.24) but, as has been seen from the account given, the inverse deduction of the principle from the equations of motion requires the assumption of a new postulate represented by (2.18). This is at variance with the view that the whole of mechanics is based on Newton's laws. The difficulty lies in the nature of constraints. For calculational purposes these are idealized to the extent that they imply the existence of discontinuous forces. Such entities cannot, of course, exist in nature; though they can be approximated to quite closely. If the idealization is regarded as a desirable one, an additional postulate, lying outside the Newtonian scheme, is necessary for their inclusion in a general description.* The term Newtonian mechanics will however be used often in a loose general sense to include d'Alembert's principle as well as Newton's laws.

d'Alembert's principle is often invoked as a means of determining the forces of constraint. Immediately, this is not possible, since the statement of the principle specifically excludes such forces. The method is to consider an alternative system in which one or more of the constraining forces is replaced by an applied force. The latter may then be calculated by making a virtual displacement of the system. The value so obtained also represents the corresponding constraint force in the original system.

One special case of (2.24) is worthy of mention. Suppose the virtual displacements coincide with an actual displacement of the system in the time dt. Then:

$$\delta' \mathbf{r}_s = \dot{\mathbf{r}}_s \, dt \tag{2.26}$$

(2.24) now becomes:

$$\sum_s \left(\mathbf{F}_s^{(a)} - \frac{d}{dt}(m_s \dot{\mathbf{r}}_s) \right) . \dot{\mathbf{r}}_s \, dt = 0$$

i.e.,

$$\sum_s \mathbf{F}_s^{(a)} . d\mathbf{r}_s - \frac{d}{dt} \left(\sum_s \tfrac{1}{2} m_s \dot{\mathbf{r}}_s^2 \right) dt = 0 \tag{2.27}$$

$\sum_s \tfrac{1}{2} m_s \dot{\mathbf{r}}_s^2$ is recognizable as the kinetic energy T of the system and,

with the further restriction that the system is a conservative one (i.e., that $\mathbf{F}_s^{(a)} = -\nabla_s V$), (2.27) may be rewritten:

$$d(V + T) = 0 \tag{2.28}$$

which states that the total energy of the system remains constant during its motion, i.e., the principle of the conservation of energy has been recovered as a special case of d'Alembert's principle.

* An alternative view is that (2.18) could be *deduced* from Newton's laws using some form of limit process. This is true of particular cases but, in the absence of a general proof, (2.18) is best regarded as an independent postulate.

CHAPTER III

The Lagrangian Formulation

The aim of the present chapter will be to provide an alternative prescription to Newton's for the writing down of the equations of motion. The guiding principles will be to base considerations on energy expressions as far as possible and to frame all equations to be equally applicable in any generalized co-ordinate system.

Conservative Systems with no Constraints

For a system of N particles there are $3N$ equations of motion of the form:

$$F_i = \frac{d}{dt}(m_i \dot{x}_i) \tag{3.1}$$

and the kinetic energy of the system is defined as:

$$T = \sum_i \tfrac{1}{2} m_i \dot{x}_i{}^2 \tag{3.2}$$

Combining these two equations gives:

$$F_i = \frac{d}{dt}\left(\frac{\partial T}{\partial \dot{x}_i}\right) \tag{3.3}$$

By definition a conservative system is one for which the force components F_i are given by:

$$F_i = -\frac{\partial V}{\partial x_i} \tag{3.4}$$

where $V = V(x_i)$ is the potential energy of the system. Hence (3.3) now becomes:

$$\frac{d}{dt}\left(\frac{\partial T}{\partial \dot{x}_i}\right) = -\frac{\partial V}{\partial x_i} \tag{3.5}$$

This equation already partially conforms to the requirements

17

stated above in that the equations of motion are expressed in terms of two scalar functions T and V. The next stage is to replace the cartesian by a generalized co-ordinate system.

Let such a system be denoted by a representative co-ordinate q_i where in general:

$$q_i \equiv q_i(x_1, x_2, \ldots x_{3N}, t) \equiv q_i(x_j, t) \qquad (i = 1, 2, \ldots 3N) \quad (3.6)$$

The inclusion of a dependence of the q_i on t is desirable since it may be required to consider moving co-ordinate systems.

The form of equation (3.5) may be regarded as suggestive of the general procedure and corresponding derivatives determined in the new co-ordinate system. As a first step:

$$\frac{\partial T}{\partial \dot{q}_i} = \sum_j m_j \dot{x}_j \frac{\partial \dot{x}_j}{\partial \dot{q}_i} \qquad (3.7)$$

the inverse * relation to (3.6) is:

$$x_j = x_j(q_i, t) \qquad (3.6')$$

hence: $\qquad \dot{x}_j \equiv \dfrac{dx_j}{dt} = \sum_i \dfrac{\partial x_j}{\partial q_i} \dot{q}_i + \dfrac{\partial x_j}{\partial t}$

from which: $\qquad \dfrac{\partial \dot{x}_j}{\partial \dot{q}_i} = \dfrac{\partial x_j}{\partial q_i} \qquad (3.8)$

thus (3.7) becomes:

$$\frac{\partial T}{\partial \dot{q}_i} = \sum_j m_j \dot{x}_j \frac{\partial x_j}{\partial q_i} \qquad (3.9)$$

Taking the time derivative of this last result gives:

$$\frac{d}{dt}\left(\frac{\partial T}{\partial \dot{q}_i}\right) = \sum_j \left\{ m_j \ddot{x}_j \frac{\partial x_j}{\partial q_i} + m_j \dot{x}_j \frac{d}{dt}\left(\frac{\partial x_j}{\partial q_i}\right) \right\}$$

---

* It is a matter of assumption that an inverse relation does exist. Mathematically this need not be so, but in all cases of physical interest the assumption is true except perhaps at a limited number of singular points.

$$= \sum_j \left\{ F_j \frac{\partial x_j}{\partial q_i} + \frac{\partial}{\partial q_i}(\tfrac{1}{2}m_j \dot{x}_j{}^2) \right\}^*$$

$$= Q_i + \frac{\partial T}{\partial q_i} \tag{3.10}$$

where:

$$Q_i \equiv \sum_j F_j \frac{\partial x_j}{\partial q_i} \tag{3.11}$$

The quantities Q_i may be termed the generalized force components. If, again, the system is a conservative one:

$$Q_i = - \sum_j \frac{\partial V}{\partial x_j} \frac{\partial x_j}{\partial q_i} = - \frac{\partial V}{\partial q_i} \tag{3.11'}$$

and (3.10) now becomes:

$$\frac{d}{dt}\left(\frac{\partial T}{\partial \dot{q}_i}\right) = \frac{\partial T}{\partial q_i} - \frac{\partial V}{\partial q_i} \tag{3.10'}$$

which contains only derivatives of the scalar quantities T and V. The generalization has brought in a term $\dfrac{\partial T}{\partial q_i}$ additional to those appearing in (3.5). This is the generalized form of such terms as centrifugal and Coriolis forces, usually referred to as fictitious forces. It may be described as arising from the curvature of the co-ordinate surfaces and is, of course, identically zero for any cartesian system.

Equation (3.10') may be written in a more compact form by the introduction of a new function L defined by:

$$L = T - V \tag{3.12}$$

this function is termed the *Lagrangian* function of the system. (3.10') now becomes:

$$\frac{d}{dt}\left(\frac{\partial L}{\partial \dot{q}_i}\right) = \frac{\partial L}{\partial q_i} \tag{3.13}$$

since by definition $\dfrac{\partial V}{\partial \dot{q}_i} = 0$.

---

* Note that:
$$\frac{d}{dt}\left(\frac{\partial x_j}{\partial q_i}\right) = \sum_k \frac{\partial}{\partial q_k}\left(\frac{\partial x_j}{\partial q_i}\right)\dot{q}_k + \frac{\partial}{\partial t}\left(\frac{\partial x_j}{\partial q_i}\right)$$
$$= \frac{\partial}{\partial q_i}\left(\sum_k \frac{\partial x_j}{\partial q_k}\dot{q}_k + \frac{\partial x_j}{\partial t}\right) = \frac{\partial \dot{x}_j}{\partial q_i}$$

The $3N$ equations (3.13) are still to be regarded as the equations of motion of the system since they represent a reformulation of (3.1). In the present form they represent a very elegant condensation of the properties of the system. It is to be noted, however, that the restriction to a conservative system is still present. The general case is represented by (3.10) which constitutes some advance on the Newton's laws formulation since the troublesome fictitious force terms are determined automatically by calculating the $\dfrac{\partial T}{\partial q_i}$. It is still necessary, however, separately to specify each component of the remaining forces.

Non-conservative Systems

Suppose that the generalized force components are given by:

$$Q_i = \frac{d}{dt}\left(\frac{\partial M}{\partial \dot{q}_i}\right) - \frac{\partial M}{\partial q_i} \tag{3.14}$$

where $M = M(q_i, \dot{q}_i, t)$ is some function of the time, the co-ordinates and their derivatives. (3.10) can then again be written in the form:

$$\frac{d}{dt}\left(\frac{\partial L}{\partial \dot{q}_i}\right) = \frac{\partial L}{\partial q_i} \tag{3.13}$$

if:
$$L = T - M \tag{3.15}$$

It might be supposed that equation (3.14) represents too severe a restriction upon the functional dependence of the force components Q_i for it to serve any useful purpose. In fact, however, it covers the highly important case of the motion of charged particles in an electromagnetic field. In vector notation the force on a particle of charge e is given (using Gaussian units) by the Lorentz expression:

$$\mathbf{F} = e\left(\mathbf{E} + \frac{1}{c}\mathbf{v} \wedge \mathbf{B}\right) \tag{3.16}$$

in this form it is not suitable for incorporation in the general scheme; it may be made so by introducing the scalar and vector potentials ϕ and $\mathbf{A}$ as alternative descriptions of the field, where:

$$\left.\begin{array}{l} \mathbf{B} = \nabla \wedge \mathbf{A} \\[2mm] \mathbf{E} = -\nabla\phi - \dfrac{1}{c}\dfrac{\partial \mathbf{A}}{\partial t} \end{array}\right\} \tag{3.17}$$

In terms of these potential functions the components of the Lorentz force become (after some reduction):

$$F_i = \left(\frac{d}{dt} \frac{\partial}{\partial \dot{x}_i} - \frac{\partial}{\partial x_i} \right) e \left(\phi - \frac{1}{c} \mathbf{v}.\mathbf{A} \right) \quad (3.18)$$

which is of the form (3.14) but is in terms of a cartesian co-ordinate system. It remains to transform to generalized co-ordinates. Substituting $M = e \left(\phi - \frac{1}{c} \mathbf{v}.\mathbf{A} \right)$ in (3.11) gives, for a representative force component in the generalized system:

$$Q_i = \sum_j F_j \frac{\partial x_j}{\partial q_i} = \sum_j \left(\frac{d}{dt} \frac{\partial M}{\partial \dot{x}_j} - \frac{\partial M}{\partial x_j} \right) \frac{\partial x_j}{\partial q_i}$$

$$= \sum_j \left\{ \frac{d}{dt} \left(\frac{\partial M}{\partial \dot{x}_j} \frac{\partial x_j}{\partial q_i} \right) - \frac{\partial M}{\partial \dot{x}_j} \frac{d}{dt} \left(\frac{\partial x_j}{\partial q_i} \right) - \frac{\partial M}{\partial x_j} \frac{\partial x_j}{\partial q_i} \right\}$$

$$= \sum_j \left\{ \frac{d}{dt} \left(\frac{\partial M}{\partial \dot{x}_j} \frac{\partial \dot{x}_j}{\partial \dot{q}_i} \right) - \left(\frac{\partial M}{\partial \dot{x}_j} \frac{\partial \dot{x}}{\partial q_i} + \frac{\partial M}{\partial x_j} \frac{\partial x_j}{\partial q_i} \right) \right\} \quad \text{[using (3.8)]}$$

$$= \frac{d}{dt} \left(\frac{\partial M}{\partial \dot{q}_i} \right) - \frac{\partial M}{\partial q_i} \quad (3.19)$$

since $\qquad M = M(\dot{x}_j, x_j) \qquad$ and $\qquad x_j = x_j(q_i, t)$.

The last result is now identical with the requirement (3.14). The Lorentz force thus conforms to this more general prescription for incorporation in the Lagrangian scheme, and the equations of motion of a particle moving in an electromagnetic field may be written in the form:

$$\frac{d}{dt} \left(\frac{\partial L}{\partial \dot{q}_i} \right) - \frac{\partial L}{\partial q_i} = 0 \quad (3.13)$$

where: $\qquad L = T - M = T - e \left(\phi - \frac{1}{c} \mathbf{v}.\mathbf{A} \right) \quad (3.20)$

Constraints

So far it has been assumed that the $3N$ position co-ordinates of a system of N particles are all independently variable. There may, however, be constraints so that the number of degrees of freedom of the system is less than $3N$. The forces required to maintain the constraints vary with the motion of the system and cannot be determined until that motion is itself known. As a result it is not easy to see how they can be included in the potential function from which, in the Lagrangian description, the forces were supposed derivable.

If the constraints are *holonomic* they may be included in the Lagrangian formalism. The procedure in this case is to invoke d'Alembert's principle. From (2.24) this principle can be expressed in the form:

$$\sum_i \left\{ F_i^{(a)} - \frac{d}{dt}(m_i \dot{x}_i) \right\} \delta x_i = 0 \qquad (3.21)$$

where the $F_i^{(a)}$ are the applied force components (i.e., *those excluding the forces of constraint*). The explicit restriction to *reversible* infinitesimal displacements has been dropped since this is implied by the proviso that the constraints be holonomic.

Since the δx_i are restricted by the conditions of constraint they are not all arbitrary and it is not possible to equate each coefficient of (3.21) to zero. It is first necessary to transform to a set of co-ordinates all of which are independently variable. This may be carried out in the following systematic manner:

Let the equations of constraint (already supposed holonomic) be r equations of the form:

$$f_l(x_i, t) = 0 \qquad (l = 1, 2, 3 \ldots r) \qquad (3.22)$$

Any $3N$ functions of the x_i and t would constitute a set of generalized co-ordinates provided they may be solved uniquely for the x_i. Choose as the first r of such a set the r functions $f_l(x_i, t)$ appearing in (3.22) and the remaining $n = (3N - r)$ in any convenient way; i.e., choose:

$$q_l(x_i, t) = f_l(x_i, t) (= 0) \qquad (l = 1, 2, 3 \ldots r) \qquad (3.23a)$$
$$q_l(x_i, t) = F_l(x_i, t) \qquad (l = (r + 1), (r + 2), \ldots 3N) \qquad (3.23b)$$

where the $F_l(x_i, t)$ are not identically zero.

Ignoring the zero conditions, equations (3.23) represent a general co-ordinate transformation between two sets of $3N$ variables. In any case of physical interest the equations may be solved giving:

$$x_i = x_i(q_j, t) \qquad (j = 1, 2, \ldots 3N) \qquad (3.24)$$

Since this is a general non-singular co-ordinate transformation the general argument of the first section may now be used to transform equation (3.21) into the form:

$$\sum_{i=1}^{3N} \left\{ Q_i^{(a)} - \frac{d}{dt}\left(\frac{\partial T}{\partial \dot{q}_i}\right) + \frac{\partial T}{\partial q_i} \right\}\delta q_i = 0 \qquad (3.25)$$

where:
$$Q_i^{(a)} = \sum_j F_j^{(a)}\frac{\partial x}{\partial q_i} \qquad (3.25')$$

Taking account now of the constraint conditions (3.22), the first r terms of the summation of (3.25) are identically zero since the corresponding δq_i's must vanish. This leaves:

$$\sum_{i=r+1}^{3N} \left\{ Q_i^{(a)} - \frac{d}{dt}\left(\frac{\partial T}{\partial \dot{q}_i}\right) + \frac{\partial T}{\partial q_i} \right\}\delta q_i = 0 \qquad (3.26)$$

Although superficially similar to (3.25) this relation is in fact profoundly different since the $n = (3N - r)$ co-ordinates q_i are now all independently variable. It follows that the coefficients of the δq_i can now be equated to zero. Hence:

$$Q_i^{(a)} - \frac{d}{dt}\left(\frac{\partial T}{\partial \dot{q}_i}\right) + \frac{\partial T}{\partial q_i} = 0 \qquad (i = (r + 1), (r + 2) \ldots 3N) \; (3.27)$$

These equations are identical in form with (3.10) and with similar limitations may be written:

$$\frac{d}{dt}\left(\frac{\partial L}{\partial \dot{q}_i}\right) = \frac{\partial L}{\partial q_i} \qquad (i = (r + 1), (r + 2) \ldots 3N) \quad (3.27')$$

It has thus been shown that the equations of motion can be written in Lagrangian form when there are holonomic constraints. Further extension is possible only to those non-holonomic cases where the constraints are expressible as non-integrable differential relations. Consideration of these will be deferred until after that of the variational principle in Chapter VI. It will then also be found possible to develop a process (method of undetermined multipliers) for determining the magnitudes of the constraint forces.

The Rayleigh Dissipation Function

In many systems there are dissipative forces present. The term covers processes, such as friction, by which energy is effectively lost to the system. A sufficiently detailed description would, in principle, identify these forces in a manner describable by the methods so far developed. However, this would involve unacceptable complications and it is convenient to treat them in a more phenomenological manner.

Dissipative forces are found experimentally in many cases to vary with the velocity components according to the simple law:

$$F_i^{(d)} = - k_i \dot{x}_i \tag{3.28}$$

where the k_i are constants. They may be incorporated in the analytical description by defining a new quantity, the *Rayleigh dissipation function*, given by

$$R = \tfrac{1}{2} \sum_i k_i \dot{x}_i^2 \tag{3.29}$$

from which:

$$F_i^{(d)} = - \frac{\partial R}{\partial \dot{x}_i} \tag{3.30}$$

Assuming that the system is otherwise describable in Lagrangian terms the equations of motion in cartesian co-ordinates become:

$$\frac{d}{dt}\left(\frac{\partial L}{\partial \dot{x}_i}\right) - \frac{\partial L}{\partial x_i} + \frac{\partial R}{\partial \dot{x}_i} = 0 \tag{3.31}$$

Transforming to a generalized co-ordinate system the generalized dissipative force components would be given by:

$$Q_i^{(d)} = \sum F_j^{(d)} \frac{\partial x_j}{\partial q_i} = - \sum_j k_j \dot{x}_j \frac{\partial x_j}{\partial q_i}$$

again, from the transformation relations:

$$\frac{\partial x_j}{\partial q_i} = \frac{\partial \dot{x}_j}{\partial \dot{q}_i} \tag{3.8}$$

$$\therefore Q_i^{(d)} = - \sum k_j \dot{x} \frac{\partial \dot{x}_j}{\partial \dot{q}_i} = \frac{\partial}{\partial \dot{q}_i}\left\{ - \tfrac{1}{2} \sum_j k_j \dot{x}_j^2 \right\} = - \frac{\partial R}{\partial \dot{q}_i} \tag{3.32}$$

Combining this result with the result for the transformation of the unmodified form of the equations of motion (i.e., excluding the dissipative forces) gives:

$$\frac{d}{dt}\left(\frac{\partial L}{\partial \dot{q}_i}\right) - \frac{\partial L}{\partial q_i} + \frac{\partial R}{\partial \dot{q}_i} = 0 \qquad (3.33)$$

It is thus seen that dissipative systems may, in suitable circumstances, be brought into an extended Lagrangian formulation.

CHAPTER IV

Applications of
Lagrange's Equations

In this chapter a number of specific problems are considered in
Lagrangian terms. Since the object of this method is to provide a
consistent way of *formulating* the equations of motion it will not be
considered necessary, in general, to deduce all the details of the
motion. The problems considered do not form a comprehensive
collection. The aim has been to provide a familiarity with the work-
ings of the Lagrangian method from a study of a few selected
examples which are of considerable interest in their own right.

A major difficulty in the solution of any problem in mechanics is
the selection of that co-ordinate system which will leave the equations of motion in a form most amenable to further treatment. This
applies equally to the Lagrangian or any other method. As a rule
there is no short cut to this selection process and it usually amounts
to one of trial and error. Space considerations naturally dictate that
any account of the failures be avoided in the problems considered.
Consequently, only those co-ordinate systems will be considered
which have, on previous experience, been found to be suitable.

A feature that should be noted is the frequent emergence of
ignorable (or *cyclic*) co-ordinates which allows a simple first integration of the corresponding equations of motion. This introduces an
important development which will be considered in greater detail in
connection with the Hamiltonian theory of the next chapter.

Coriolis and Centrifugal Forces
The introduction of a rotating co-ordinate system often complicates
rather than simplifies a mechanical problem though, as in the case of
motion of a body in the earth's atmosphere, it may be a necessary
consideration.

Consider a transformation from a set of stationary axes $Oxyz$ to a set $Ox'y'z'$ rotating with angular velocity ω about Oz. The equations of transformation are:

$$\left. \begin{array}{l} x = x' \cos \omega t - y' \sin \omega t \\ y = x' \sin \omega t + y' \cos \omega t \\ z = z' \end{array} \right\} \tag{4.1}$$

The kinetic energy of a particle is now given by:

$$\begin{aligned} T &= \tfrac{1}{2}m(\dot{x}^2 + \dot{y}^2 + \dot{z}^2) \\ &= \tfrac{1}{2}m(\dot{x}'^2 + \dot{y}'^2 + \dot{z}'^2) + m\omega(x'\dot{y}' - \dot{x}'y') + \tfrac{1}{2}m\omega^2(x'^2 + y'^2) \end{aligned} \tag{4.2}$$

It was previously stated that the term $\dfrac{\partial T}{\partial q_i}$ appearing in the Lagrangian equations of motion may be referred to as a fictitious force arising from peculiarities of the co-ordinate system. In the present case this takes the form:

$$\frac{\partial T}{\partial x'} = m\omega\dot{y}' + m\omega^2 x' \tag{4.3}$$

with a corresponding expression for $\dfrac{\partial T}{\partial y'}$ $\left(\dfrac{\partial T}{\partial z'} = 0 \right)$.

The two terms of (4.3) are identifiable as components of half the Coriolis force and of the centrifugal force respectively. The remaining half of the Coriolis force component derives from the term $\dfrac{d}{dt}\left(\dfrac{\partial T}{\partial \dot{x}} \right)$ of Lagrange's equations.

It is not intended to consider any specific problem involving Coriolis or centrifugal forces, the aim is merely to demonstrate how simply they emerge from the Lagrangian formulation. This is in contrast to the alternative vector method, which is equally valid in principle, but which often presents practical difficulties, particularly with regard to sign.

The Two-Body Problem

In terms of their position vectors $\mathbf{r}_1$ and $\mathbf{r}_2$ the motion of a system consisting of two particles, acted upon only by conservative forces, can be described by a Lagrangian function.

$$L = T - V = \tfrac{1}{2}m_1\dot{\mathbf{r}}_1^2 + \tfrac{1}{2}m_2\dot{\mathbf{r}}_2^2 - V(\mathbf{r}_1, \mathbf{r}_2) \tag{4.4}$$

The kinetic energy T may alternatively be expressed as:

$$T = \tfrac{1}{2}\mu\dot{\mathbf{r}}^2 + \tfrac{1}{2}M\dot{\mathbf{R}}^2 \tag{4.5}$$

where $\mathbf{r} = \mathbf{r}_2 - \mathbf{r}_1$ ($=$ vector separation of the particles)

$\mathbf{R} = \dfrac{m_1\mathbf{r}_1 + m_2\mathbf{r}_2}{m_1 + m_2}$ ($=$ position of centre of mass of the system)

$\mu = \dfrac{m_1 m_2}{m_1 + m_2}$ ($=$ *reduced* mass of the system)

$M = m_1 + m_2$

$$\tag{4.6}$$

In this alternative mode of description it appears that the system consists of one particle of mass equal to the sum of the masses of the original particles and situated at the position of the centre of mass, together with a second one having the reduced mass of the system and lying at the point $\mathbf{r} = (\mathbf{r}_2 - \mathbf{r}_1)$. On the face of it this is an artificial alternative. Its point lies in the fact that, with commonly realized restrictions, the potential energy term of the Lagrangian splits into two terms which correlate with those in (4.5). We have then:

$$V = V_1(\mathbf{r}) + V_2(\mathbf{R}) \tag{4.7}$$

Assuming this, the Lagrangian may be written as:

$$L = L_1 + L_2 \tag{4.8}$$

where
$$\begin{aligned} L_1 &= \tfrac{1}{2}\mu\dot{\mathbf{r}}^2 - V_1(\mathbf{r}) \\ L_2 &= \tfrac{1}{2}M\dot{\mathbf{R}}^2 - V_2(\mathbf{R}) \end{aligned} \tag{4.9}$$

and the motion of the original system of two interacting particles has been effectively decomposed into the independent motions of two independent systems. Such a decomposition normally simplifies the solution of the problem.

The separation represented by (4.7) occurs either if there are no external forces acting on the system or if the external force per unit mass is the same for each particle. As previously mentioned, these restrictions are realized sufficiently often for it to be worth while to consider the problem in this form.

It is of little interest here to consider further that part of the motion of the system represented by L_2. It is, however, proposed to say more about the other component with the additional restriction that:

$$V_1(\mathbf{r}) = V_1(r) \qquad (r = |\mathbf{r}|) \tag{4.10}$$

i.e., that the motion takes place under a *central force* law.

Using *spherical polar co-ordinates* defined in the usual way:

$$\dot{\mathbf{r}}^2 = (\dot{r}^2 + r^2\dot{\theta}^2 + r^2 \sin^2\theta\dot{\phi}^2) \tag{4.11}$$

hence: $$L_1 = \tfrac{1}{2}\mu(\dot{r}^2 + r^2\dot{\theta}^2 + r^2 \sin^2\theta\dot{\phi}^2) - V_1(r) \tag{4.12}$$

It is seen that ϕ is here an *ignorable co-ordinate*, i.e., it does not appear explicitly in the expression for L_1. As a consequence the corresponding Lagrange's equation $\left(\dfrac{d}{dt}\dfrac{\partial L}{\partial \dot{\phi}} = \dfrac{\partial L}{\partial \phi}\right)$ takes the particularly simple form:

$$\frac{d}{dt}(\mu r^2 \sin^2\theta\dot{\phi}) = 0 \tag{4.13}$$

An immediate integration gives:

$$\mu r^2 \sin^2\theta\dot{\phi} = \text{const.} \tag{4.14}$$

this quantity may be identified as the angular momentum about the polar axis and is seen to be constant for any particular choice of that axis. If the axis is chosen initially to lie along $\mathbf{r}$, then $\theta = 0$ initially, i.e., the value of the constant in (4.14) is zero. In the course of the subsequent motion r and θ will, in general, assume non-zero values. It follows that $\dot{\phi}$ must vanish in order that (4.14) can continue to be satisfied. This implies that the polar axis and the direction of motion remain coplanar throughout the motion. As a consequence it will simplify matters to reframe the problem in terms of a *plane* polar co-ordinate system, the plane in question being that containing $\mathbf{r}$ and $\dot{\mathbf{r}}$. In this way the Lagrangian becomes:

$$L_1 = \tfrac{1}{2}\mu(\dot{r}^2 + r^2\dot{\theta}^2) - V_1(r) \tag{4.15}$$

In this new formulation θ is now seen to be an ignorable co-ordinate and the solution to the corresponding equation of motion is:

$$\mu r^2\dot{\theta} = \text{const.} \ (= l) \tag{4.16}$$

This can be interpreted as stating the constancy of the angular momentum about an axis through the origin and normal to the co-ordinate plane. It is essentially Kepler's second law of planetary motion and is a consequence of the assumption of central forces.

The equation of motion relating to the r co-ordinate is less simple:

$$\mu\ddot{r} = \mu r\dot{\theta}^2 - \frac{\partial}{\partial r}V_1(r) = \frac{l^2}{\mu r^3} - \frac{\partial}{\partial r}V_1(r) = -\frac{\partial}{\partial r}\left(\frac{l^2}{2\mu r^2} + V_1(r)\right) \quad (4.17)$$

A useful interpretation is that of a particle moving in one dimension under the influence of a modified potential function given by:

$$V_1'(r) = \frac{l^2}{2\mu r^2} + V_1(r) \quad (4.18)$$

Further progress requires an assumption concerning the specific form of $V_1(r)$.

The above reasoning demonstrates the importance of the choice of co-ordinate system in bringing out those features of the problem which can be summarized as conservation laws. Although these laws do not constitute the whole solution to the problem they do form an essential part of it.

Larmor's Theorem

The motion of an electron in the field of a positively charged nucleus can be considered as a classical two-body problem. Since the external forces are zero it fulfils the conditions of the previous example and the motion is separable into the two independent components. The introduction of an external magnetic field alters this state of affairs since the external force per unit mass is then not the same for each particle, and it is not possible to express the external forces as a function only of **R**, the position of the centre of mass of the two particles.

However, bearing in mind the much greater mass of the nucleus, it is a good approximation to regard the position of the latter as fixed and to consider only the motion of the electron. The problem is then essentially a one-body problem. From the considerations of the last chapter the effect of the magnetic field on the electron is represented by a velocity dependent term $-\frac{e}{c}\mathbf{v}.\mathbf{A}$ appearing in the Lagrangian

and the coulomb attraction of the nucleus is represented by
$V(r) = Ze^2/r$; thus:

$$L = T - V(r) - \frac{e}{c}\mathbf{v}.\mathbf{A} \tag{4.19}$$

Assuming a constant magnetic field of magnitude H_0 parallel to
the z-axis, a possible set of cartesian components for A are:

$$A_x = -\tfrac{1}{2}H_0y \qquad A_y = \tfrac{1}{2}H_0x \qquad A_z = 0 \tag{4.20}$$

Cylindrical polar co-ordinates are found to be more suited to the
solution of this problem. Transforming to these we have:

$$A_\rho = A_z = 0 \qquad A_\theta = \tfrac{1}{2}H_0\rho \tag{4.20'}$$

and equation (4.19) becomes:

$$L = \tfrac{1}{2}m(\dot{\rho}^2 + \rho^2\dot{\theta}^2 + \dot{z}^2) - V(\rho, z) - \frac{e}{c}\rho\dot{\theta}\tfrac{1}{2}H_0\rho \tag{4.19'}$$

Where θ is seen to be an ignorable co-ordinate and the integration
of the corresponding equation of motion gives:

$$m\rho^2\left(\dot{\theta} - \frac{eH_0}{2mc}\right) = \text{const.} \tag{4.21}$$

As will be seen from later considerations, this result again represents
the conservation of angular momentum, though at present only the
term $m\rho^2\dot{\theta}$ can be recognized as a quantity of that sort. For present
purposes it is convenient to consider the co-ordinate transformation:

$$\rho = \rho', \qquad z = z', \qquad \theta = \alpha + \omega_0t \quad \text{where} \quad \omega_0 = \frac{eH_0}{2mc} \tag{4.22}$$

The transformed Lagrangian is:

$$L' = \tfrac{1}{2}m(\dot{\rho}'^2 + \rho'^2\dot{\alpha}^2 + \dot{z}'^2) - V(\rho', z') - \tfrac{1}{2}m\rho'^2\omega_0^2 \tag{4.23}$$

Also, in the absence of the magnetic field H_0, the original Lagrangian
takes the form:

$$L = \tfrac{1}{2}m(\dot{\rho}^2 + \rho^2\dot{\theta}^2 + \dot{z}^2) - V(\rho, z) \tag{4.24}$$

(4.23) and (4.24) are of identical form but for the term $-\tfrac{1}{2}m\rho'^2\omega_0^2$.
A relevant consideration is the ratio of the magnitude of this term
to that of the quantity $\tfrac{1}{2}m\rho^2\dot{\theta}^2$ and it may be shown that this is of
the order of 10^{-12} for values of H_0 as large as 10^4 oersted. To a very

high order of approximation, therefore, the term may be ignored and the Lagrangian functions (4.23) and (4.24) are identical in form. It follows that the motions of the systems described by the two Lagrangians are the same. Physically this means that, under the influence of the uniform magnetic field, the motion of the electron relative to the rotating co-ordinate system is identical with what it would be in a stationary co-ordinate system with no applied field. The result is usually expressed by saying that the system precesses with constant angular velocity ω_0 when under the influence of the field.

It should be noted that it has only really been shown that the possible states of motion of the two systems are the same. However, it can also be proved that, if a steady magnetic field is *gradually* built up, the system retains its state of motion relative to a co-ordinate system rotating with the appropriate angular velocity.

The result may be extended to multi-electron systems in which the electrons move in a central force field (i.e., due to a single nucleus). It is of wide application in the atomic theory of the magnetic properties of substances.*

The Symmetrical Top

The spinning-top is an example of the motion of a rigid body. The system is thus one in which holonomic time-independent constraints reduce the maximum number of degrees of freedom to six; in cases of interest this number is further reduced to three by requiring the peg to be in contact with the ground at a fixed point. Ignoring any frictional forces which may act at the peg the system is a conservative one in which the only applied force is the weight of the top acting through its centre of gravity.

The constraints are automatically allowed for by expressing the motion in terms of the moments and products of inertia of the top. These are quasi-geometrical quantities, and if full advantage is to be taken of the symmetry of the top its axis of symmetry must coincide with one axis of a cartesian co-ordinate system. The products of inertia are then zero and are thus effectively removed from the considerations. The axis of symmetry will, in general, be moving in space

* For further discussion of the theorem see *Theory of Electrons* by L. Rosenfeld (N. Holland Publishing Co., Amsterdam, 1951).

and it will further be necessary to relate the moving co-ordinate system to a stationary one. A convenient set of parameters for describing the system is known as the *Eulerian angles*. These are three in number: θ the angle between the axis of symmetry of the top (Oz') and a fixed vertical axis (Oz), ψ the angle between Ox (fixed in space) and Ox' the line of intersection of a plane perpendicular to Oz' and

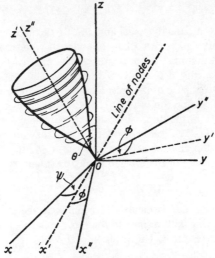

FIG. 1 The symmetrical top

a horizontal plane (Ox' is known as the *line of nodes*), ϕ the angle between Ox' and Ox'', where the latter is an axis fixed with respect to the top and perpendicular to Oz' ($\equiv Oz''$).

The axes $Ox'y'z'$ are rotating with respect to $Oxyz$ with an instantaneous angular velocity $\mathbf{\Omega}'$. Measured in the $Ox'y'z'$ system:

$$\mathbf{\Omega}' = (\dot{\theta}, \ \dot{\psi} \sin \theta, \ \dot{\psi} \cos \theta) \tag{4.25}$$

The axes $Ox''y''z''$ (with respect to which the top is stationary) are rotating with an angular velocity $\mathbf{\omega}$ with respect to $Ox'y'z'$. Again measured in the $Ox'y'z'$ system:

$$\mathbf{\omega} = (0, \ 0, \ \dot{\phi}) \tag{4.26}$$

The top is thus rotating, with respect to the stationary system $Oxyz$, with an angular velocity $\boldsymbol{\Omega} = \boldsymbol{\Omega}' + \boldsymbol{\omega}$. Measured in the system $Ox'y'z'$:

$$\boldsymbol{\Omega} = (\dot{\theta},\ \dot{\psi}\sin\theta,\ \dot{\phi} + \dot{\psi}\cos\theta) \qquad (4.27)$$

With respect to a general set of cartesian axes the kinetic energy of the top is:

$$T = \tfrac{1}{2}(A\Omega_x^2 + B\Omega_y^2 + C\Omega_z^2 - 2D\Omega_y\Omega_z - 2E\Omega_z\Omega_x - 2F\Omega_x\Omega_y) \qquad (4.28)$$

where A, B, C are the moments and D, E, F the products of inertia. The top is varying its orientation continuously with respect to $Oxyz$; hence, if the latter are used as reference axes A, B, C, D, E, F as well as the components of $\boldsymbol{\Omega}$ will vary with time. This would constitute a gross inconvenience in the calculation which can be overcome by referring the motion to $Ox''y''z''$. These axes are not only stationary with respect to the top, so that the corresponding quantities A'', B'', C'', D'', E'', F'' are constant in time, they are also principal axes for the top and the products of inertia D'', E'', F'' all vanish. The kinetic energy of the top can then be written:

$$T = \tfrac{1}{2}(A''\Omega_{x''}^2 + B''\Omega_{y''}^2 + C''\Omega_{z''}^2) \qquad (4.29)$$

Finally, since Oz'' is an axis of symmetry, the system $Ox'y'z'$, although moving with respect to the top, similarly constitutes a set of principal axes for it. This gives:

$$T = \tfrac{1}{2}(A'\Omega_{x'}^2 + A'\Omega_{y'}^2 + C'\Omega_{z'}^2) \qquad (4.29')$$

where $A' = A'' = B''$, $C' = C''$.

In view of (4.27) this may also be written:

$$T = \tfrac{1}{2}\Big(A'\dot{\theta}^2 + A'\dot{\psi}^2\sin^2\theta + C'(\dot{\psi}\cos\theta + \dot{\phi})^2\Big) \qquad (4.30)$$

(Note that this gives the kinetic energy of the top *measured with respect to a stationary frame of reference* but using a moving coordinate system. The kinetic energy of the top *with respect to the moving frame of reference* represented by $Ox'y'z'$ would be $\tfrac{1}{2}C'\dot{\phi}^2$.)

The potential energy of the top is:

$$V = mgh\cos\theta \qquad (4.31)$$

where h is the distance of the centre of gravity from the peg. Hence the motion of the top is described by a Lagrangian function:

$$L = T - V$$
$$= \tfrac{1}{2}\left(A'\dot\theta^2 + A'\dot\psi^2 \sin^2\theta + C'(\dot\psi \cos\theta + \dot\phi)^2\right) - mgh \cos\theta \quad (4.32)$$

This completes the translation of the physical circumstances of the problem into mathematical terms. It now remains to derive the equations of motion according to the standard prescription of Lagrange's equations. The parameters θ, ψ, ϕ are obvious choices as generalized co-ordinates and it is apparent that both ϕ and ψ are ignorable. The equations of motion corresponding to these ignorable co-ordinates integrate immediately giving:

$$\left.\begin{array}{l} (\dot\psi \cos\theta + \dot\phi) = \text{const.} = n \\ A'\dot\psi \sin^2\theta + C' \cos\theta(\dot\psi \cos\theta + \dot\phi) \\ \qquad = A'\dot\psi \sin^2\theta + C'n \cos\theta = \text{const.} = k \end{array}\right\} \quad (4.33)$$

The remaining equation of motion is:

$$A'\ddot\theta = A'\dot\psi^2 \sin\theta \cos\theta - C'n\dot\psi \sin\theta + mgh \sin\theta \quad (4.34)$$

Further progress with the solution requires information concerning initial conditions and is not our present concern. The general discussion of the motion is noted for its complexity but the treatment above has formulated the problem with the minimum of effort. The equations (4.33) again represent conservation of angular momenta though the components concerned are physically not so important as in previous cases.

Normal Modes of Vibration

Certain physical systems undergo a restricted form of motion consisting of small displacements about a position of stable equilibrium. An example of such a motion is the mechanical vibration of an atomic lattice as in a crystal. The motion is complicated but can be resolved into a linear combination of a finite number of simple harmonic motions. In general, each component S.H.M. represents a motion of the whole lattice. The components are known as the *normal modes of vibration* of the system.

Consider a system of mutually interacting linearly vibrating particles. With the postulate that the motion of the system is of small amplitude and takes place about a position of stable equilibrium the potential energy of the system may be expressed as a Taylor's series:

$$V(q_i) = V(q_i^{(0)}) + \sum_i \delta q_i \left(\frac{\partial V}{\partial q_i}\right)$$
$$+ \sum_i \sum_j \tfrac{1}{2} \delta q_i \delta q_j \left(\frac{\partial^2 V}{\partial q_i \partial q_j}\right)_0 + \dots \quad (4.35)$$

where q_i are the generalized co-ordinates of the particles and $q_i^{(0)}$ the values of these co-ordinates at equilibrium.

For the purposes of the problem it is the δq_i's rather than the q_i's which are of importance. We thus re-define:

$$\delta q_i = \eta_i \quad (4.36)$$

and treat these as generalized co-ordinates. Since $\frac{\partial}{\partial q_i} \equiv \frac{\partial}{\partial \eta_i}$ equation (4.35) may be rewritten as:

$$V(\eta_i) = V(0) + \sum_i \eta_i \left(\frac{\partial V}{\partial \eta_i}\right)_0 + \sum_i \sum_j \tfrac{1}{2} \eta_i \eta_j \left(\frac{\partial^2 V}{\partial \eta_i \partial \eta_j}\right)_0 + \dots \quad (4.35')$$

$V(0)$ is an arbitrary constant which may be treated as zero, also $\left(\frac{\partial V}{\partial \eta_i}\right)_0 = 0$ since, by hypothesis, $\eta_i = 0$ is a position of equilibrium. (4.35') thus becomes:

$$V(\eta_i) = \sum_i \sum_j \tfrac{1}{2} \eta_i \eta_j V_{ij} + 0(\eta^3) \quad (4.35'')$$

where the $V_{ij} \equiv \left(\frac{\partial^2 V}{\partial \eta_i \partial \eta_j}\right)_0$ are independent of the η_i.

In terms of a cartesian co-ordinate system the kinetic energy is given by:

$$T = \sum_i \tfrac{1}{2} m_i \dot{x}_i^2 \quad (4.37)$$

where the m_i are the masses of the individual particles. Transforming to the generalized system via the transformation relations:

$$q_i = q_i(x_i) \quad \text{or} \quad x_j = x_j(q_i) \quad (4.38)$$

where *time dependence is explicitly excluded*, gives, remembering that $q_i = q_i^{(0)} + \eta_i$:

$$\dot{x}_i = \sum_j \frac{\partial x_i}{\partial q_j} \dot{q}_j = \sum_j \frac{\partial x_i}{\partial \eta_j} \dot{\eta}_j \qquad (4.39)$$

hence:

$$T = \sum_i \tfrac{1}{2} m_i \left(\sum_j \frac{\partial x_i}{\partial \eta_j} \dot{\eta}_j \right)^2 = \sum_i \tfrac{1}{2} m_i \left(\sum_j \sum_k \frac{\partial x_i}{\partial \eta_j} \frac{\partial x_i}{\partial \eta_k} \dot{\eta}_j \dot{\eta}_k \right) \qquad (4.40)$$

again assuming the amplitude of motion is small we may put:

$$\frac{\partial x_i}{\partial \eta_j} = \left(\frac{\partial x_i}{\partial \eta_j} \right)_0 + \sum_k \eta_k \left(\frac{\partial^2 x_i}{\partial \eta_k \partial \eta_j} \right)_0 + \ldots \qquad (4.41)$$

thus:

$$T = \sum_j \sum_k \tfrac{1}{2} M_{jk} \dot{\eta}_j \dot{\eta}_k + 0(\eta \dot{\eta}^2) \qquad (4.42)$$

where

$$M_{jk} \equiv \sum_i m_i \left(\frac{\partial x_i}{\partial \eta_j} \right)_0 \left(\frac{\partial x_i}{\partial \eta_k} \right)_0 \qquad (4.43)$$

is independent of the η_i and t.

Neglecting the higher order quantities the Lagrangian of the system (assumed conservative) may now be written:

$$L = T - V = \sum_j \sum_k \tfrac{1}{2} \{ M_{jk} \dot{\eta}_j \dot{\eta}_k - V_{jk} \eta_j \eta_k \} \qquad (4.44)$$

and the equations of motion are:

$$\frac{d}{dt} \sum_k (M_{jk} \dot{\eta}_k) = - \sum_k V_{jk} \eta_k \qquad (4.45)$$

i.e.,

$$\sum_k M_{jk} \ddot{\eta}_k = - \sum_k V_{jk} \eta_k \qquad (4.46)$$

From these it is seen that there is a coupling between the motions of the particles. The assumption is now made that the motion is periodic, i.e.,

$$\eta_k = \eta_k^{(0)} e^{i\omega t} \qquad (4.47)$$

substituting in (4.46) gives:

$$\sum_k (\omega^2 M_{jk} - V_{jk}) \eta_k^{(0)} = 0 \qquad (4.48)$$

this is a set of $3N$ equations linking the $3N$ quantities $\eta_k^{(0)}$. The condition for the existence of a non-trivial solution is:

$$| \omega^2 M_{jk} - V_{jk} | = 0 \qquad (4.49)$$

where the left-hand side represents the determinant whose (j, k)th element is $(\omega^2 M_{jk} - V_{jk})$.

The solution of this determinantal equation provides $3N$ values of ω^2 corresponding to the $3N$ frequencies of the normal modes of vibration of the system. The $3N$ solutions are all independent and a general motion of the system consists of an arbitrary linear combination of them. It is to be emphasized that particular modes are not, as a rule, associated with individual particles. In general, the motion of each particle has a component at each of the normal frequencies. Some of the values of ω^2 may be negative, ω is then purely imaginary and represents an unstable component of the motion. The aperiodic components are still sometimes referred to as modes of vibration in the general sense, though their consideration is really ruled out by the initial stipulation that the system moves about a position of stable equilibrium.

The effect of including higher order terms in the Lagrangian (4.44) is, strictly, to exclude the possibility of the resolution of the motion into independent modes. The determination of such generalized *anharmonic* motion is a difficult process. However, it is usual to treat the higher order terms as second order effects which cause interactions between the normal modes. The general motion of the system can then once more be represented as a linear combination of the $3N$ harmonic terms, with initially arbitrary coefficients. Due to the interactions the coefficients will now vary with time in a manner determined by the higher order 'anharmonic' terms.

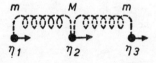

The generalities of the problem of the determination of normal modes are well illustrated by a simple classical model which gives a good representation of the behaviour of a linear triatomic molecular system. In this model a particle of mass M is elastically coupled to two other particles each of mass m. The elastic constant is μ in each case and in the equilibrium position the particles are equally spaced on a straight line. Only the motion along the line is to be considered.

If the displacements of the particles from equilibrium are denoted by η_1, η_2, η_3, the kinetic energy is given by:

$$T = \tfrac{1}{2}m(\dot{\eta}_1{}^2 + \dot{\eta}_3{}^2) + \tfrac{1}{2}M\dot{\eta}_2{}^2 \qquad (4.50)$$

also

$$V = \tfrac{1}{2}\mu(\eta_3 - \eta_2)^2 + \tfrac{1}{2}\mu(\eta_2 - \eta_1)^2 \qquad (4.51)$$

$$\therefore \ L = \tfrac{1}{2}m(\dot{\eta}_1{}^2 + \dot{\eta}_3{}^2) + \tfrac{1}{2}M\dot{\eta}_2{}^2 - \tfrac{1}{2}\mu(\eta_3 - \eta_2)^2 - \tfrac{1}{2}\mu(\eta_2 - \eta_1)^2 \quad (4.52)$$

the equations of motion are thus

$$\left.\begin{aligned}
m\ddot{\eta}_1 &= \mu(\eta_2 - \eta_1) \\
M\ddot{\eta}_2 &= \mu(\eta_3 - \eta_2) - \mu(\eta_2 - \eta_1) = \mu(\eta_3 + \eta_1 - 2\eta_2) \\
m\ddot{\eta}_3 &= -\mu(\eta_3 - \eta_2)
\end{aligned}\right\} \qquad (4.53)$$

Assuming S.H.M. gives:

$$\left.\begin{aligned}
(m\omega^2 - \mu)\eta_1{}^0 + \mu\eta_2{}^0 &= 0 \\
(M\omega^2 - 2\mu)\eta_2{}^0 + \mu(\eta_3{}^0 + \eta_1{}^0) &= 0 \\
(m\omega^2 - \mu)\eta_3{}^0 + \mu\eta_2{}^0 &= 0
\end{aligned}\right\} \qquad (4.54)$$

the determinantal equation is thus :

$$\begin{vmatrix}
m\omega^2 - \mu & \mu & 0 \\
\mu & M\omega^2 - 2\mu & \mu \\
0 & \mu & m\omega^2 - \mu
\end{vmatrix} = 0 \qquad (4.55)$$

solving this equation gives

$$\omega^2 = 0, \ \mu/m, \ \mu(2m + M)/mM \qquad (4.56)$$

The normal frequencies are thus

$$\omega_1 = 0, \qquad \omega_2 = \pm\sqrt{\mu/m}, \qquad \omega_3 = \pm\sqrt{\frac{\mu(2m + M)}{mM}} \qquad (4.57)$$

As in Fourier analysis, there is no reality to negative frequencies. The assumed solution was of exponential form; the pairs of such solutions with equal and opposite values of ω combine to give a single *cos* or *sin* solution, the two arbitrary constants reappearing as arbitrary amplitude and phase values.

The solution $\omega_1 = 0$ corresponds to the physically possible solution in which the three particles undergo simultaneously the same translatory motions. The solution $\omega_2 = \sqrt{\mu/m}$ gives $\eta_2{}^0 = 0$ and

$\eta_1^0 = -\eta_3^0$, corresponding to a motion in which the centre particle is stationary and the outer ones move in antiphase. The third solution represents a motion with the outer particles moving in phase with each other and in antiphase with the centre one.

As in the general case, it would be possible to determine the explicit form of a transformation $\eta_i' = \eta_i'(\eta_j)$ to a new co-ordinate system in which each mode is associated with only one co-ordinate. It will be assumed here that the physically significant information lies in a knowledge of the normal frequencies and no attempt will be made to discover the required transformation. This is commonly the case, though sometimes it may be necessary to solve the problem completely.

The above considerations illustrate the power of the Lagrangian method in discussing problems concerning small vibrations in general terms. The solution of the determinantal equation in a problem involving a large number of degrees of freedom (as with a crystal lattice) may be very difficult, but the above method can always be used as a starting-point.

Electrical Circuits

It is interesting, but not particularly useful from a practical point of view, to note that electrical circuit analysis can be formulated in Lagrangian terms. Consider the Lagrangian function:

$$L = \tfrac{1}{2} \sum_i \sum L_{ij} \dot{I}_i \dot{I}_i + \sum_i \dot{E}_i I_i - \tfrac{1}{2} \sum_i \frac{1}{C_i} I_i^2 \qquad (4.58)$$

where $L_{ij} = L_{ji}$.

The related equations of motion would be:

$$\frac{d}{dt}\left(\sum L_{ij} \dot{I}_j \right) = \dot{E}_i - I_i/C_i$$

i.e.,
$$\dot{E}_i = I_i/C_i + \sum_j L_{ij} \ddot{I}_j \qquad (4.59)$$

If the L_{ij} $(i \neq j)$ are interpreted as mutual inductances, L_{ij} $(i = j)$ as self-inductances and the C_i as capacitances, these equations are the correct circuital relations for a network of coupled reactive impedances in which are flowing a system of currents I_i generated by e.m.f.s E_i. It is clear that such problems can be formulated in analytical terms using the Lagrangian function given by equation (4.58) and with the mesh currents used as generalized co-ordinates. (Resistive impedances may also be incorporated in the formalism by using the Rayleigh dissipation function.)

The analogy between mechanical and electrical systems is usually developed by noting the similarity in form of the equations of motion.* At this level it is of great importance. Methods devised for the solution of specialized electric circuit problems are often taken over and re-applied to the solution of the equivalent mechanical problems. The reverse process rarely occurs in practice because of the greater effort which has, in the past, been devoted to electrical problems. The similarity in the Lagrangian formalisms merely reflects this correspondence between the equations of motion and is unlikely of itself to lead to further results. The usefulness of the Lagrangian formulation lies, in general, in providing a convenient method for deriving the equations of motion and this rarely proves difficult in electric circuit analysis.

* See, for example: *Wave Propagation in Periodic Structures* by L. Brillouin (Dover Publications Inc., 1953).

CHAPTER V

The Hamiltonian Formulation

Momentum

It was seen in considering applications of the Lagrangian method
that ignorable co-ordinates imply the constancy of quantities which
are sometimes identifiable, from previous knowledge, as momenta.
It should, however, be firmly emphasized that the term momentum
nowhere appears explicitly in connection with the Lagrangian treatment. It is an essential feature of the formalism that the independent
variables are the time and the generalized co-ordinates. The time
derivatives of the generalized co-ordinates also *appear* explicitly in
the formulation, but it is always understood that they are to be
treated ultimately as dependent variables. This fact is illustrated by
the use of the concept of a trajectory in configuration space to
represent the motion of a system.

In line with the use of generalized co-ordinates it is possible to
introduce generalized momenta. Their use, however, takes us outside the Lagrangian scheme and into a new type of description
usually associated with Hamilton.

Generalized momentum components are defined by:

$$p_i = \frac{\partial L}{\partial \dot{q}_i} \tag{5.1}$$

there is thus one component for each generalized co-ordinate and
each q_i, p_i combination is said to form a pair of *conjugate* variables.
There is no obvious justification for the definition at this stage apart
from noting that it leads, in the case of conservative systems, to those
quantities familiarly regarded as linear and angular momentum
components. On the other hand, the alternative definition $p_i = \dfrac{\partial T}{\partial \dot{q}_i}$
would also do this. The real justification lies in the general consistency
and success of the theory which can be based on the definition given.

With non-conservative systems the general definition may embrace quantities not normally recognized as momenta. Consider the case of a charged particle moving in an electromagnetic field. This is an example of a non-conservative system describable in Lagrangian terms. From Chapter III the Lagrangian function is:

$$L = T - e \left(\phi - \frac{1}{c} \, \mathbf{v} . \mathbf{A} \right) \tag{5.2}$$

From the definition (5.1), the momentum components in a cartesian system will be given by:

$$p_i = m\dot{x}_i + eA_i/c \tag{5.3}$$

These include, in addition to the familiar $m\dot{x}_i$ terms, a new one, eA_i/c, which is usually referred to as the *electromagnetic momentum*.

In terms of the definition (5.1) Lagrange's equations of motion become

$$\frac{\partial L}{\partial q_i} = \frac{d}{dt} p_i \tag{5.4}$$

which bear some resemblance to the original Newtonian form of the equations of motion.

Ignorable Co-ordinates

The definition of momentum components leads to the generalization that if:

$$\frac{\partial L}{\partial q_i} = 0 \tag{5.5}$$

then

$$\frac{\partial L}{\partial \dot{q}_i} = \text{const.} \tag{5.6}$$

i.e.,

$$p_i = \text{const.} \tag{5.6'}$$

Ignorable co-ordinates thus always imply the constancy of the corresponding momentum components. The constancy of linear and angular momenta in conservative systems provides a special case of the general rule. In the consideration of Larmor's theorem it was found that the effect of a magnetic field on a monatomic system was a general precession of the system about the direction of the field. An alternative way of looking at the problem is that the momentum conjugate to the angular co-ordinate θ is conserved on applying the field, the increase in the electromagnetic momentum being compensated by a reduction in the mechanical part.

Phase Space

The introduction of momentum components is used to effect a complete change in viewpoint. As was stated above, the Lagrangian method regards the co-ordinates of a system as the independent quantities specifying the state of a system. Each of these variables is obtained as a function of time from the solution of the set of second order differential equations known as Lagrange's equations. An alternative approach is to regard both co-ordinates and momenta as independent variables. The final aim of any problem is then to obtain all of these quantities as explicit functions of time.

There is no evident advantage to be gained from this change of attitude. It would not, for instance, have assisted in the solution of any of the problems of Chapter IV to have substituted the symbol p_i for the quantities $\frac{\partial L}{\partial \dot{q}_i}$. A method for solving the modified form of the equations of motion which appear in the new formulation can be developed and may occasionally have advantages over that for solving the Lagrangian equations (see Chapter VII). This is not likely to occur often, however, and it must be admitted that a knowledge of the Lagrangian suffices for tackling most ordinary problems in mechanics. The real gain of the alternative formulation is that it provides a convenient basis for the development of quantum mechanics and of statistical mechanics. Although only quantum mechanics will be considered explicitly in this book it is to be understood that a great deal of what follows is spade work in preparation for an appreciation of the formulation of both subjects.

Configuration space was introduced as a verbal device for displaying the motion of a system when employing the Lagrangian approach. This is no longer adequate when momenta as well as space co-ordinates are regarded as independently variable. Instead, the history of the system may be regarded as being represented by a path in a *phase space* of $6N$ dimensions, each particle contributing one dimension for each position co-ordinate and one for each momentum component. As was pointed out in connection with configuration space, the geometrical language is illustrative only; any conceptual difficulties may at once be removed by substitution of the word 'variable' for 'dimension'.

A great deal of mental difficulty can be experienced in attempting to reconcile the configuration space and phase space concepts as true alternatives. Some insight into this problem may be gained by realizing that there is essentially more arbitrariness about the path in configuration space than the one in phase space. The specification of the form of the Lagrangian function (equivalent to fixing the equations of motion) is the starting-point in either case, but it does not fix the path. If, in addition to the Lagrangian function, one point of the path of the system in phase space is given, then the whole path is determined, since stipulating one point is, in effect, fixing six initial values for each particle. On the other hand, a single point in configuration space only supplies three initial values for each particle and other information must be given to fix the path. Put another way, if the equations of motion are defined, there are an infinite number of possible paths through any point in configuration space but only one possible path in phase space.

The Hamiltonian Function

The Lagrangian function is a function, in general, of the q_i, the $\dot{q}_i$ and t. Its total time derivative is therefore given by:

$$\frac{dL}{dt} = \sum_i \frac{\partial L}{\partial q_i}\dot{q}_i + \sum_i \frac{\partial L}{\partial \dot{q}_i}\ddot{q}_i + \frac{\partial L}{\partial t} \tag{5.7}$$

Using the Lagrangian equations of motion (3.13) this becomes:

$$\frac{dL}{dt} = \sum_i \frac{d}{dt}\left(\frac{\partial L}{\partial \dot{q}_i}\right)\dot{q}_i + \sum_i \frac{\partial L}{\partial \dot{q}_i}\ddot{q}_i + \frac{\partial L}{\partial t} = \sum_i \frac{d}{dt}\left(\dot{q}_i\frac{\partial L}{\partial \dot{q}_i}\right) + \frac{\partial L}{\partial t} \tag{5.8}$$

i.e.,
$$\frac{d}{dt}\left\{\sum_i \dot{q}_i\frac{\partial L}{\partial \dot{q}_i} - L\right\} = -\frac{\partial L}{\partial t} \tag{5.8'}$$

If
$$\frac{\partial L}{\partial t} = 0 \quad \text{then} \quad \left\{\sum_i \dot{q}_i\frac{\partial L}{\partial \dot{q}_i} - L\right\} = \text{const.} \tag{5.9}$$

or, using the definition (5.1):

$$\left\{ \sum_i p_i \dot{q}_i - L \right\} = \text{const.} \tag{5.9'}$$

We now make the definition

$$H = \sum_i p_i \dot{q}_i - L \tag{5.10}$$

introducing a new function, *the Hamiltonian*, which can be seen from dimensional considerations to be (like L) an energy function. Part of its importance can be deduced from the fact that, as shown above, it is a constant of motion if t does not appear explicitly in the Lagrangian. It will later appear that H is to be identified with the total energy in most cases of physical interest.

A further aspect of the significance of the Hamiltonian function can be seen from the following considerations. For some purposes it is preferable to frame analytical mechanics in terms of the variables q_i, p_i and t rather than q_i, $\dot{q}_i$ and t. A similar situation occurs in thermodynamics where, when it is desired to switch from entropy S and volume V to temperature T and V as independent variables, a new energy function F (the Helmholtz free energy) is derived from the previous one U (the internal energy) through the defining relation:

$$F = U - TS$$

F is then regarded as a function of T and V, whereas U was a function of S and V. This is an example of a Legendre transformation. The definition (5.10) is seen to be another example of the same procedure and it follows that the Hamiltonian function is to be regarded, in general, as a function of the position co-ordinates, the momenta and the time.

i.e., $$H = H(q_i, p_i, t) \tag{5.11}$$

As obtained using the definition (5.10) in connection with a given Lagrangian function, H will appear as a function of the q_i, the $\dot{q}_i$ and t. It is understood, however, that equations (5.1) are to be solved for the $\dot{q}_i$ as functions of q_i, p_i and t and the results substituted in (5.10) giving the required functional dependence.

Alternative forms of the equations of motion can be derived using the Hamiltonian function. From (5.11) it follows that:

$$dH = \sum_i \frac{\partial H}{\partial q_i} dq_i + \sum_i \frac{\partial H}{\partial p_i} dp_i + \frac{\partial H}{\partial t} dt \qquad (5.12)$$

also from (5.10)

$$dH = \sum_i \dot{q}_i \, dp_i + \sum_i p_i \, d\dot{q}_i - dL \qquad (5.13)$$

however, $\qquad L = L(q_i, \dot{q}_i, t)$

$$\therefore dL = \sum_i \frac{\partial L}{\partial q_i} dq_i + \sum_i \frac{\partial L}{\partial \dot{q}_i} d\dot{q}_i + \frac{\partial L}{\partial t} dt$$

$$= \sum_i \frac{\partial L}{\partial q_i} dq_i + \sum_i p_i d\dot{q}_i + \frac{\partial L}{\partial t} dt \text{ [using (5.1)] } \qquad (5.14)$$

hence combining (5.13) and (5.14):

$$dH = - \sum_i \frac{\partial L}{\partial q_i} dq_i + \sum_i \dot{q}_i dp_i - \frac{\partial L}{\partial t} dt$$

$$= - \sum_i \dot{p}_i dq_i + \sum_i \dot{q}_i dp_i - \frac{\partial L}{\partial t} dt \text{ from (5.4) } \qquad (5.15)$$

comparison of the coefficients in (5.12) and (5.15) now shows:

$$\dot{q}_i = \frac{\partial H}{\partial p_i} \qquad \dot{p}_i = - \frac{\partial H}{\partial q_i} \qquad (5.16)$$

$$\frac{\partial L}{\partial t} = - \frac{\partial H}{\partial t} \qquad (5.17)$$

(5.16) are referred to as *Hamilton's canonical form of the equations of motion*. In principle they constitute an advance in formulation since they are first order differential equations whereas Lagrange's equations are of second order. In practice this gain is largely illusory. The simplest condition for the solubility of any one of the equations is that a q_i or a p_i does not appear explicitly in the Hamiltonian; the corresponding conjugate quantity is then a constant of the motion. The solution of the equations thus reduces to the problem of the determination of a co-ordinate system in which sufficient of the q_i's and p_i's are ignorable. This can be reduced to a recipe (the

Hamilton–Jacobi method to be studied in Chapter VII). Unfortunately, however, it involves the solution of a first order partial differential equation which can be of a type presenting great difficulties. It can be carried through in certain cases, but the main reason for studying the Hamiltonian formulation is, as previously stated, as a convenient basis for quantum and statistical mechanics.

Physical Significance of the Hamiltonian

When encountering applications of the Hamiltonian formulation for the first time the beginner often receives the impression that the Hamiltonian function is a pointless synonym for the total energy of the system under consideration. As previously noted it does, like the Lagrangian also, have the dimensions of energy and in most useful cases it reduces to the total energy; nevertheless it is not identical with this quantity in all circumstances. The conditions for equality will now be investigated.

For a conservative system $\dfrac{\partial V}{\partial \dot{q}_i} = 0$ by definition, hence:

$$p_i = \frac{\partial L}{\partial \dot{q}_i} = \frac{\partial T}{\partial \dot{q}_i} \tag{5.18}$$

now:

$$T = \sum_i \tfrac{1}{2} m_i \dot{x}_i^2 \tag{5.19}$$

and in general: $x_i = x_i(q_j, t)$ \hfill (5.20)

hence

$$\dot{x}_i = \sum_j \frac{\partial x_i}{\partial q_j} \dot{q}_j + \frac{\partial x_i}{\partial t} \tag{5.21}$$

and:

$$T = \sum_j \sum_k a_{jk} \dot{q}_j \dot{q}_k + \sum b_j \dot{q} + c \tag{5.22}$$

where:

$$
\left.
\begin{aligned}
a_{jk} &= \sum \tfrac{1}{2} m_i \frac{\partial x_i}{\partial q_j} \frac{\partial x_i}{\partial q_k} \\[2mm]
b_j &= \sum_i m_i \frac{\partial x_i}{\partial q_j} \frac{\partial x_i}{\partial t} \\[2mm]
c &= \sum_i \tfrac{1}{2} m_i \left(\frac{\partial x_i}{\partial t} \right)^2
\end{aligned}
\right\} \tag{5.23}
$$

It follows that T is a homogeneous quadratic form in the generalized velocities if t does not appear explicitly in the transformation relations (5.20), i.e., under these conditions:

$$T = \sum_j \sum_k a_{jk}\dot{q}_j\dot{q}_k \tag{5.24}$$

and it may then easily be verified* that:

$$\sum_i \dot{q}_i\frac{\partial T}{\partial \dot{q}_i} = 2T \tag{5.25}$$

it follows in this case that:

$$H = \sum_i p_i\dot{q}_i - L = T + V = E\,(= \text{total energy}) \tag{5.26}$$

Note that the restrictions made were (a) the system is a conservative one, (b) the co-ordinate transformation is independent of time (i.e., the axes are fixed in space). These are sufficient but not necessary conditions for the equality of E and H. Another case for which it holds is that of a charged particle moving in an unchanging electromagnetic field. For such a system it was previously found:

$$L = T - e\left(\phi - \frac{1}{c}\mathbf{v}.\mathbf{A}\right) \tag{5.2}$$

$$p_i = m\dot{x}_i + eA_i/c \tag{5.3}$$

hence: $$\sum_i p_i\dot{q}_i = 2T + \frac{e}{c}\mathbf{v}.\mathbf{A} \tag{5.27}$$

and: $$H = T + e\phi \tag{5.28}$$

The latter quantity may also be identified as the total energy of the system from a consideration of the line integral of the force over the path of the particle as in Chapter II. In this case the equality of H and E occurs partly because of a seemingly fortuitous cancellation of terms involving the vector potential. This may be traced further to the fact that the velocity dependent potential terms of the Lagrangian are homogeneous linear functions of the velocity components. If

---

* This result is a special case of Euler's theorem which states that $\sum_i x_i\frac{\partial f}{\partial x_i} = nf$ if the function f is a homogeneous n-ic. in the x_i.

these terms are denoted by $L^{(v)}$, then it follows from Euler's theorem that $\sum_i \dot{x}_i \dfrac{\partial L^{(v}}{\partial \dot{x}_i} - L^{(v)} = 0$. The non-appearance of $L^{(v)}$ in the expression for H is required if the latter is to represent the total energy, since the forces derived from the velocity dependent potential terms do no work in the course of the motion. In general it may be assumed that H and E will be identical unless the co-ordinate system is moving with respect to the frame of reference. It must, of course, be admitted that the principal virtue of the Hamiltonian is that it provides a method for the determination of the energy which, like the whole Lagrangian and Hamiltonian schemes, does not require the individual identification of force components.

The constancy of H in time is a separate matter. Since $H = H(p_i, q_i, t)$ it follows that:

$$\frac{dH}{dt} = \sum_i \frac{\partial H}{\partial q_i}\dot{q}_i + \sum_i \frac{\partial H}{\partial p_i}\dot{p}_i + \frac{\partial H}{\partial t} = \frac{\partial H}{\partial t} \qquad (5.29)$$

by virtue of the canonical relations (5.16). From the similarity of this result with the second of those relations it would appear that H bears the same relationship to $-t$ as does p_i to q_i, i.e., H and $-t$ may be regarded as conjugate variables. The analogy between (5.29) and the first of (5.16) would suggest a different correspondence which, however, is not confirmed by relativistic considerations (see Chapter X).

Combining (5.17) and (5.29) gives:

$$\frac{dH}{dt} = \frac{\partial H}{\partial t} = - \frac{\partial L}{\partial t} \qquad (5.30)$$

The condition for H to be a constant of motion is thus that t shall not appear explicitly in L. This condition is necessary as well as sufficient.

The greatest interest lies in systems for which the total energy is a constant of motion. This usually involves a simultaneous satisfaction of the condition for $E = H$ and for $H = $ const., but again this is not a necessary requirement as may be seen from the following example.

Consider a frictionless sleeping-top (i.e., one which is spinning about its vertical axis of symmetry). This is an example of a system in which E is conserved. Using the symbols of the example in Chapter IV:

$$T = \tfrac{1}{2}c\phi^2 \qquad V = mgh$$
$$\therefore L = T - V = \tfrac{1}{2}c\phi^2 - mgh$$

transforming to a new co-ordinate α given by $\alpha = (\phi - \omega_0 t)$, it follows that

$$\dot{\alpha} = \phi - \omega_0, \quad L = \tfrac{1}{2}c(\dot{\alpha} + \omega_0)^2 - mgh \quad \text{and} \quad p_\alpha = \frac{\partial L}{\partial \dot{\alpha}} = c(\dot{\alpha} + \omega_0)$$

hence: $$H = (p_\alpha \dot{\alpha} - L) = \left(\frac{p_\alpha^2}{2c} - \omega_0 p_\alpha + mgh\right)$$

whereas: $$E = T + V = \left(\frac{p_\alpha^2}{2c} + mgh\right) \neq H.$$

This is a somewhat trivial case but serves to show that H and E may both be constants of motion without being identical. The unusual character of H here arises from the use of a co-ordinate system which is rotating with respect to the frame of reference and it cannot be identified with the energy with respect either to the stationary or the moving system.

Constants of Motion and Symmetry

Mention has already been made of circumstances under which the Hamiltonian function and the momentum co-ordinates remain constant during the motion of the system. From one point of view the constancy of a momentum component is an incidental aspect of the fact that a co-ordinate is ignorable, the main outcome being that the corresponding equation of motion (either Lagrangian or Hamiltonian) is immediately integrable. From another point of view the constancy itself is regarded as an important property of the system. The alternative view looms large in more advanced applications of the formalism in modern physics and it may even constitute an acceptable solution to a problem to identify all the constants of motion. In the general sense the term *constant of motion* is applied to any dynamical variable and not merely to the Hamiltonian and the momenta. Methods of identifying these constants generally will form an important part of later considerations (see Chapter VIII).

If a cartesian co-ordinate is ignorable it follows that the Hamiltonian and Lagrangian functions are invariant with respect to a translation of the system along the corresponding axis. An ignorable angular co-ordinate likewise implies invariance with respect to rotation. Since these ignorable co-ordinates lead to constancy of the corresponding momenta it follows that constants of motion are associated with symmetry aspects of the system. In view of (5.29) there is a similar symmetry relation between the Hamiltonian and the time co-ordinate. In general, conservation and symmetry properties are so related that the terms are used almost interchangeably.

Dissipative Systems

In Chapter III it was shown that dissipative systems may be brought within a modified Lagrangian scheme by the introduction of a new function, the Rayleigh dissipation function, in addition to the Lagrangian function itself.

This new function was defined as:

$$R = \tfrac{1}{2} \sum_j k_j \dot{x}_j^2 \tag{3.29}$$

and the modified equations of motion in terms of generalized co-ordinates are:

$$\frac{d}{dt} \frac{\partial L}{\partial \dot{q}_i} - \frac{\partial L}{\partial q_i} + \frac{\partial R}{\partial \dot{q}_i} = 0 \tag{3.31}$$

Such a description does not lead to a useful formulation in the Hamiltonian sense since the energy of the system is not constant and cannot be identified with the Hamiltonian in any way. There is an alternative approach which goes some way to meeting this difficulty. It has as its basic idea the extension of the dissipative system under consideration to include another, similar but hypothetical, system in which the dissipated energy is absorbed. This is a purely mathematical device but gives a combined system in which the total energy is conserved. As an example consider the case of the damped linear simple harmonic oscillator with an equation of motion:

$$m\ddot{x} + k\dot{x} + \mu x = 0 \qquad (5.31)$$

where k is the damping constant.

The equation of motion of the complementary (or 'mirror image') system will likewise be:

$$m\ddot{x}' - k\dot{x}' + \mu x' = 0 \qquad (5.32)$$

Consider the function:

$$L = m(\dot{x}\dot{x}') - \tfrac{1}{2}k(x'\dot{x} - \dot{x}'x) - \mu xx' - E_0 \qquad (5.33)$$

where E_0 is the initial energy calculated from assumed initial conditions.

This may be regarded as a Lagrangian function for the combined system described by the variables x and x' since equations (5.31) and (5.32) are reproduced by the application of the standard rules. The corresponding momenta are:

$$\left.\begin{aligned}
p &= \frac{\partial L}{\partial \dot{x}} = m\dot{x}' - \tfrac{1}{2}kx' \\
p' &= \frac{\partial L}{\partial \dot{x}'} = m\dot{x} + \tfrac{1}{2}kx
\end{aligned}\right\} \qquad (5.34)$$

and the Hamiltonian function:

$$H = \sum_i p_i\dot{q}_i - L = m\dot{x}\dot{x}' + \mu xx' + E_0 \qquad (5.35a)$$

$$= \frac{1}{m}(p + \tfrac{1}{2}kx')(p' - \tfrac{1}{2}kx) + \mu xx' + E_0 \qquad (5.35b)$$

Since $\dfrac{\partial H}{\partial t} = 0$ it follows that $H = $ const. This is also seen from the realization that the solutions to the equations of motion will take the form:

$$\left.\begin{aligned}
x &= x_0 e^{i\omega t}e^{-\alpha t} \\
x' &= x_0' e^{i\omega t}e^{+\alpha t}
\end{aligned}\right\} \qquad (5.36)$$

where $\omega = \sqrt{\dfrac{\mu}{m} - \dfrac{k^2}{4m^2}}$, $\alpha = \dfrac{k}{2m}$. Substitution in (5.35a) yields the result $H = E_0$, confirming the identity of the Hamiltonian and the total energy. It is not clear that the mathematical device serves any useful purpose in the case of discrete systems but it can be utilized in the consideration of continuous ones. Note that the conjugate variables p and p' have lost all physical significance in this new approach.

CHAPTER VI

Variational Principles

The variational principles which have been proposed in physics are many and varied. In some cases they have been surrounded by a philosophic mysticism which has delayed an appreciation of their value. The chief value of such principles lies in their extreme economy of expression. Here we shall consider in detail only Hamilton's principle and give a brief outline of the principle of least action.

Calculus of Variations

As a preliminary we shall consider the purely mathematical problem of determining the conditions under which a certain type of integral assumes a stationary value.

Consider the integral:

$$I = \int_{x_1}^{x_2} F(y, y', x)dx \tag{6.1}$$

where $y' \equiv \dfrac{dy}{dx}$ and it is understood that x is an independent and y a dependent variable, though the form of the dependence of y on x is not specified initially. The values of x and y at the limits of the integration are given and the value of I depends on the precise path of integration between these end points. The problem is to determine the condition under which I has a stationary value.

Let APB in fig. 2 be the path for which I is stationary and consider a neighbouring path $AP'B$ with the same end-points A, B. The correspondence between the points of the two paths is that $P \rightarrow P'$ where $P = (x, y)$ and $P' = (x, y + \delta y)$; i.e., the x coordinates of the points remain fixed. This defines a so-called δ variation of the path. Subject only to the limitation that:

$$\delta y_1 = \delta y_2 = 0 \tag{6.2}$$

the variation is an arbitrary but small one. It may otherwise be expressed as:

$$\delta y = \eta \delta \alpha \qquad (6.3)$$

where α is a parameter common to all points of the path and η is any function of x, subject to the condition:

$$\eta(x_1) = \eta(x_2) = 0 \qquad (6.2')$$

The corresponding variation in y' is:

$$\delta y' = \eta' \delta \alpha \qquad (6.4)$$

Since the variation is stipulated to be small, the integral over the

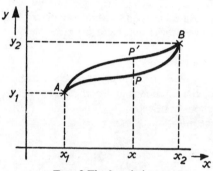

Fig. 2 The δ variation

varied path may be obtained by considering only first order terms in a Taylor series expansion of the function F:

$$I' = \int_{x_1}^{x_2} \left\{ F(y, y', x) + \frac{\partial F}{\partial y} \eta \delta \alpha + \frac{\partial F}{\partial y'} \eta' \delta \alpha \right\} dx$$

hence:

$$\delta I = \delta \alpha \int_{x_1}^{x_2} \left\{ \frac{\partial F}{\partial y} \eta + \frac{\partial F}{\partial y'} \eta' \right\} dx \qquad (6.5)$$

integrating by parts and using (6.2'):

$$\int_{x_1}^{x_2} \frac{\partial F}{\partial y'} \eta' \, dx = - \int_{x_1}^{x_2} \frac{d}{dx} \left(\frac{\partial F}{\partial y'} \right) \eta \, dx$$

$$\therefore \quad \delta I = \delta \alpha \int_{x_1}^{x_2} \left\{ \frac{\partial F}{\partial y} - \frac{d}{dx} \left(\frac{\partial F}{\partial y'} \right) \right\} \eta \, dx \qquad (6.6)$$

The condition that I is to have a stationary value implies that δI is zero. Since η is arbitrary, this in turn implies that the integrand of (6.6) must vanish

i.e.,
$$\frac{\partial F}{\partial y} - \frac{d}{dx}\left(\frac{\partial F}{\partial y'}\right) = 0 \qquad (6.7)$$

The result may readily be generalized to the case where there are n dependent variables y_i. This gives n conditions of the form:

$$\frac{\partial F}{\partial y_i} - \frac{d}{dx}\left(\frac{\partial F}{\partial y_i'}\right) = 0 \qquad (6.8)$$

If the n dependent variables y_i are functions of m independent variables x_r the n conditions become:

$$\frac{\partial F}{\partial y_i} - \sum_{r=1}^{m} \frac{d}{dx_r}\left(\frac{\partial F}{\partial y_{i,r}}\right) = 0 \qquad (6.9)$$

where $y_{i,r} \equiv \dfrac{dy_i}{dx_r}$. The mathematical statement of the principle in this case takes the form:

$$\delta \int \ldots \int F\, dx_1\, dx_2 \ldots dx_m = 0 \qquad (6.10)$$

The further generalization to include a dependence of F on higher derivatives of the y_i is also possible but has no bearing on the considerations here.

The results may be applied to a wide range of physical problems. In most cases the stationary value of the integral turns out to be a minimum one though occasionally it is a maximum. The earliest application was by Bernoulli to the determination of the path for which the time of fall of a particle between two points not in vertical line is a minimum.

In the derivation given above the conditions (6.7)–(6.9) have been shown to be consequences of the stipulation that the corresponding integral has a zero variation. They are thus necessary conditions; it can further be shown that they are sufficient ones so that if the conditions hold then it follows that the variation of the integral must be zero.

Hamilton's Principle

The equations of motion of a system described by a Lagrangian function have been shown to take the form:

$$\frac{\partial L}{\partial q_i} - \frac{d}{dt}\left(\frac{\partial L}{\partial \dot{q}_i}\right) = 0 \tag{3.13}$$

According to the last paragraph of the previous section these imply:

$$\delta \int_{t_1}^{t_2} L(q_i, \dot{q}_i, t)dt = 0 \tag{6.11}$$

This is a statement of *Hamilton's principle*. In our derivation it represents a deduction ultimately from Newton's laws. An alternative view is to regard it as a true principle in which case Lagrange's equations of motion and the rest of mechanics stem from it.

It should be emphasized that Hamilton's principle contains no more information than is already available. It merely provides a more elegant and concise formulation than the alternative postulates. It does have the advantage that it can be applied to non-mechanical systems in which, for instance, Newton's laws have no obvious meaning. This greater generality, which is an additional recommendation for adopting it as a basic postulate, will be investigated further in later chapters.

The integral:

$$S = \int_{t_1}^{t_2} L \, dt \tag{6.12}$$

where the integration is over the actual path of the system is known as *Hamilton's principal function*. Its evaluation requires a knowledge both of the path and of the Lagrangian function. The variational principle enables the former to be deduced but it is, of course, necessary to postulate the analytic form of L. To this extent Hamilton's principle is somewhat artificial. Its usefulness in new situations depends on the fact that Lagrangian functions are often quite simple functions of the possible variables.

Modified Hamilton's Principle

It was shown in Chapter V that the Lagrangian equations of motion may be replaced by a set of first order differential equations (Hamilton's canonical equations). This equivalence is borne out by the fact that the latter may also be derived from Hamilton's principle by a small modification of the argument.

From (5.10):
$$L = \sum_i p_i \dot{q}_i - H$$

where:
$$p_i = \frac{\partial L}{\partial \dot{q}_i} \qquad (5.1)$$

A modified version of Hamilton's principle thus states:

$$\delta \int_{t_1}^{t_2} \left\{ \sum_i p_i \dot{q}_i - H \right\} dt = 0 \qquad (6.13)$$

In (6.11) the variation of path allowed for variations in the q_i at constant t, the variations in the $\dot{q}_i$ were dependent on those of the q_i. In the present case, in accordance with the general assumption behind the use of the Hamiltonian function, the δ variation comprises independent variations of both the q_i and the p_i at constant t. As before, these can be expressed in terms of a parameter α common to all points of the path of integration (the latter is now a path in phase space rather than in configuration space).

i.e.,
$$\delta q_i = \frac{\partial q_i}{\partial \alpha} \delta \alpha = \eta_i \, \delta \alpha \qquad \delta p_i = \frac{\partial p_i}{\partial \alpha} \delta \alpha = \zeta_i \, \delta \alpha \qquad (6.14)$$

where the η_i and ζ_i are arbitrary subject to the conditions:

$$\eta_i(t_1) = \eta_i(t_2) = \zeta_i(t_1) = \zeta_i(t_2) = 0 \qquad (6.14')$$

The variation of the integral is thus given by:

$$\delta S = \delta \alpha \int_{t_1}^{t_2} \sum_i \left\{ \frac{\partial p_i}{\partial \alpha} \dot{q}_i + p_i \frac{\partial \dot{q}_i}{\partial \alpha} - \frac{\partial H}{\partial p_i} \frac{\partial p_i}{\partial \alpha} - \frac{\partial H}{\partial q_i} \frac{\partial q_i}{\partial \alpha} \right\} dt \quad (6.15)$$

integrating by parts and using (6.14) and (6.14'):

$$\int_{t_1}^{t_2} p_i \frac{\partial \dot{q}_i}{\partial \alpha} dt = \int_{t_1}^{t_2} p_i \frac{d}{dt} \left(\frac{\partial q_i}{\partial \alpha} \right) dt = - \int_{t_1}^{t_2} \dot{p}_i \eta_i \, dt \qquad (6.16)$$

Combining (6.15) and (6.16) gives:

$$\delta S = \delta \alpha \int_{t_1}^{t_2} \sum_i \left\{ \left(\dot{q}_i - \frac{\partial H}{\partial p_i} \right) \zeta_i - \left(\dot{p}_i + \frac{\partial H}{\partial q_i} \right) \eta_i \right\} dt \qquad (6.17)$$

The η_i and ζ_i are all independent arbitrary functions, hence the vanishing of the variation of S implies that the coefficients of these quantities are zero:

i.e.,
$$\dot{q}_i = \frac{\partial H}{\partial p_i} \qquad \dot{p}_i = -\frac{\partial H}{\partial q_i} \tag{6.18}$$

which are identical with (5.16) and thus identifiable as the canonical equations of motion.

Non-holonomic Systems

The preceding derivations assumed that all the generalized co-ordinates q_i were independent. This presumes that any constraints present are holonomic and that the co-ordinates have been chosen to correspond with the number of degrees of freedom and so automatically to include the constraints.

The equations of motion for a limited class of non-holonomic systems may also be obtained from Hamilton's principle (using *Lagrange's method of undetermined multipliers*). The class concerned is the one in which the constraint conditions are given in the form of non-integrable differential relations involving the space and time co-ordinates.

Consider such a system in which the constraint conditions are m equations of the form:

$$\sum_{i=1}^{n} a_{ri}\, dq_i + b_{rt}\, dt = 0 \qquad (r = 1, 2, \ldots m) \tag{6.19}$$

where the generalized co-ordinates are n in number and the number of degrees of freedom is $(n - m)$. The a_{ri} and b_{rt} are, in general, functions of t and of the q_i.

It is assumed that there is a Lagrangian function L, involving the time, all n co-ordinates q_i and their time derivatives, which is such that the motion of the system makes:

$$\delta \int_{t_1}^{t_2} L\, dt = 0$$

This assumption of Hamilton's principle can be developed as before to give:

$$\delta \alpha \int_{t_1}^{t_2} \sum_{i=1}^{n} \left\{ \frac{\partial L}{\partial q_i} - \frac{d}{dt}\left(\frac{\partial L}{\partial \dot{q}_i}\right) \right\} \eta_i\, dt = 0 \tag{6.20}$$

So far no account has been taken of the constraints. Their effect is that the η_i are not all independent [they are in fact restricted by the conditions (6.19)]. To make any further deductions it is necessary to reduce the number of variables to the number of degrees of freedom.

The variations of the q_i considered in developing (6.20) (δ variations) are required to occur at constant t; the restrictions on these variations due to the constraints are therefore given by the m conditions:

$$\sum_i a_{ri} \, \delta q_i = \sum_i a_{ri} \eta_i \, \delta\alpha = 0 \qquad (r = 1, 2, \ldots m) \quad (6.19')$$

These conditions can be combined with (6.20) by multiplying each of them by a (so far undetermined) quantity λ_r, integrating with respect to t over the range $t_1 - t_2$ and adding the result to (6.20). The λ_r are permitted to be functions of t but not of the other variables. (6.19') are identities, hence each of the new integrals is separately zero and the process yields:

$$\delta\alpha \int_{t_1}^{t_2} \sum_{i=1}^{n} \left\{ \frac{\partial L}{\partial q_i} - \frac{d}{dt}\left(\frac{\partial L}{\partial \dot{q}_i}\right) + \sum_{r=1}^{m} \lambda_r a_{ri} \right\} \eta_i \, dt = 0 \qquad (6.21)$$

The quantities λ_r are so far unspecified. They may therefore be chosen in such a way as to make:

$$\frac{\partial \dot{L}}{\partial q_i} - \frac{d}{dt}\left(\frac{\partial L}{\partial \dot{q}_i}\right) + \sum_{r=1}^{m} \lambda_r a_{ri} = 0 \qquad (6.22)$$

for $i = 1, 2, \ldots m$.

This choice modifies (6.21) to:

$$\delta\alpha \int_{t_1}^{t_2} \sum_{i=m+1}^{n} \left\{ \frac{\partial L}{\partial q_i} - \frac{d}{dt}\left(\frac{\partial L}{\partial \dot{q}_i}\right) + \sum_{r=1}^{m} \lambda_r a_{ri} \right\} \eta_i \, dt = 0 \qquad (6.21')$$

in which all the η_i are now independent since there are just $(n - m)$ degrees of freedom. The coefficients of the η_i may therefore be equated to zero giving the identical result to (6.22) but applying this time to $i = (m + 1), (m + 2), \ldots n$. Combining these results, equation (6.22) now holds for all the q_i. The method has incorporated the effect of the constraints in a completely symmetrical manner which does not differentiate between the co-ordinates. It is true that m of the equations were arrived at using one consideration and the remainder using another, but the results are identical in form and there is thus no effective discrimination between the various variables q_i. The final determination of the motion consists in solving for the $(n + m)$ unknown quantities q_i and λ_r, each as a function of the time. Available for this purpose are the n equations of the form (6.22) and the m equations of constraint (6.19). Usually there is no interest in the values of the λ_r and it suffices merely to eliminate them from the equations and to solve for the q_i.

The exceptional case, when it is of use to determine the λ_r, is when the magnitudes of the forces of constraint are required.

If the constraints were removed and replaced by generalized force components Q_i, the motion of the system would be described in terms of the same n co-ordinates q_i, now all independent and obeying the n equations of motion:

$$\frac{\partial L}{\partial q_i} - \frac{d}{dt}\left(\frac{\partial L}{\partial \dot{q}_i}\right) = Q_i \qquad (6.23)$$

If it be stipulated further that the motion is to coincide with that under the constraints, then (6.23) must be identical with (6.22)

i.e., $$Q_i = -\sum_{r=1}^{m} \lambda_r a_{ri} \qquad (6.24)$$

In order to produce a given motion the forces must be the same whether they are to be termed 'applied' forces or 'constraint' forces. It follows, therefore, that (6.24) gives the magnitudes of the forces of constraint.

It should be mentioned that Lagrange's method of undetermined multipliers can also be used in conjunction with d'Alembert's principle to deduce the modified equations of motion in Newtonian form.

Example

The previous section can be illustrated by the problem of determining the motion of a flat uniform disk which rolls upright without slipping on a horizontal plane. It is the 'without slipping' stipulation that requires non-holonomic constraint conditions. The rest of the constraints (rigid body, plane of disk vertical and horizontal plane of rolling) may be automatically allowed for by a suitable choice of co-ordinate system.

Ignoring the non-holonomic constraints, the system may be described by the specification of four independent co-ordinates: x, y cartesian co-ordinates representing the position of the point of contact of the disk with the horizontal plane, ψ the angle between the

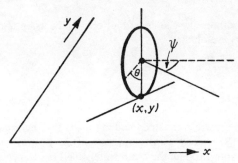

FIG. 3 Disk rolling on a horizontal plane

axis of the disk and the x axis and θ representing the angle between a fixed radius of the disk and the vertical direction. The Lagrangian describing the system is

$$L = \tfrac{1}{2}M(\dot{x}^2 + \dot{y}^2) + \tfrac{1}{2}A\dot{\theta}^2 + \tfrac{1}{2}C\dot{\psi}^2 \qquad (6.25)$$

where M is the mass of the disk, A and C its moments of inertia about its own axis and a perpendicular direction respectively.

The non-holonomic constraint can now be specified through the relation:

$$v = a\dot{\theta} \qquad (6.26)$$

where $v_x = \dot{x}$, $v_y = \dot{y}$ and a is the radius of the disk.

(6.26) is equivalent to the two conditions:

$$dx - a \sin \psi \, d\theta = 0 \qquad dy + a \cos \psi \, d\theta = 0 \qquad (6.26')$$

Hamilton's principle:

$$\delta \int L \, dt = 0$$

may be developed as before to give:

$$\delta\alpha \int \sum_i \left(\frac{\partial L}{\partial q_i} - \frac{d}{dt}\frac{\partial L}{\partial \dot{q}_i}\right)\eta_i \, dt = 0$$

i.e., $\qquad \delta\alpha \int \left\{ M\ddot{x}\eta_x + M\ddot{y}\eta_y + A\ddot{\theta}\eta_\theta + C\ddot{\psi}\eta_\psi \right\} dt = 0 \qquad (6.27)$

hence, incorporating the constraint conditions using undetermined multipliers, we obtain:

$$\delta\alpha \int \left\{ (M\ddot{x} + \lambda_1)\eta_x + (M\ddot{y} + \lambda_2)\eta_y \right.$$
$$\left. + (A\ddot{\theta} - \lambda_1 a \sin\psi + \lambda_2 a \cos\psi)\eta_\theta + C\ddot{\psi}\eta_\psi \right\} dt = 0 \qquad (6.28)$$

Two of the terms in the integrand are made zero by suitable choice of the λ's. There now remain only two of the arbitrary quantities η under the integral sign; these can be regarded as independently variable since there are just two degrees of freedom. The vanishing of the integral thus requires the coefficients of these quantities to vanish. The final result is:

$$\left. \begin{array}{l} M\ddot{x} + \lambda_1 = 0 \\ M\ddot{y} + \lambda_2 = 0 \\ A\ddot{\theta} - \lambda_1 a \sin\psi + \lambda_2 a \cos\psi = 0 \\ C\ddot{\psi} = 0 \end{array} \right\} \qquad (6.29)$$

There are six quantities $(x, y, \theta, \psi, \lambda_1, \lambda_2)$ to be determined and this is done by elimination from (6.26′) and (6.29). The final solution requires the specification of the initial values of any two of the variables x, y, θ, ψ, together with their time derivatives. The frictional forces giving rise to the constraint can be deduced from the values of λ_1 and λ_2.

The example is not quite general since terms in dt are missing from equations (6.26′). This makes no effective difference to the procedure.

However, when such terms are present, it must be remembered that their coefficients do not appear in (6.28).

Principle of Least Action

We now introduce a new and more general type of variation of the path of a system. This will be termed a Δ variation and will allow time as well as position co-ordinates to vary. At the ends of the path the position co-ordinates are held fixed but changes in the time are permitted. A point P on the unvaried path now goes over into P' on the varied path with the correspondence:

$$q_i \rightarrow q_i' = q_i + \Delta q_i = q_i + \delta q_i + \dot{q}_i \Delta t \qquad (6.30)$$

where the δ variation has the same meaning as previously and $\Delta q_i = 0$ at the end-points of the path.

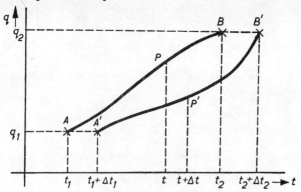

Fig. 4 Illustrating the Δ variation

The Δ variation of any function $f = f(q_i, \dot{q}_i, t)$ is given by:

$$\Delta f = \sum_i \left(\frac{\partial f}{\partial q_i} \Delta q_i + \frac{\partial f}{\partial \dot{q}_i} \Delta \dot{q}_i \right) + \frac{\partial f}{\partial t} \Delta t$$

$$= \sum_i \frac{\partial f}{\partial q_i} (\delta q_i + \dot{q}_i \Delta t) + \sum_i \frac{\partial f}{\partial \dot{q}_i} (\delta \dot{q}_i + \ddot{q}_i \Delta t) + \frac{\partial f}{\partial t} \Delta t$$

$$= \delta f + \frac{df}{dt} \Delta t \qquad (6.31)$$

i.e.

$$\Delta \equiv \delta + \Delta t \frac{d}{dt} \qquad (6.31')$$

Consider a Δ variation of Hamilton's principal function:

$$\Delta S = \Delta \int_1^2 L\,dt = \delta \int_1^2 L\,dt + \left[L\Delta t\right]_1^2$$

$$= \int_1^2 \sum_i \left(\frac{\partial L}{\partial q_i}\delta q_i + \frac{\partial L}{\partial \dot{q}_i}\delta \dot{q}_i\right)dt + \left[L\Delta t\right]_1^2 \qquad (6.32)$$

Here $\delta \int_1^2 L\,dt$ does not vanish since $\delta q_i \neq 0$ at the end-points of the path. From Lagrange's equations of motion and the relation $\delta \dot{q}_i = \frac{d}{dt}(\delta q_i)$:

$$\frac{\partial L}{\partial q_i}\delta q_i + \frac{\partial L}{\partial \dot{q}_i}\delta \dot{q}_i = \frac{d}{dt}\left(\frac{\partial L}{\partial \dot{q}_i}\right)\delta q_i + \frac{\partial L}{\partial \dot{q}_i}\frac{d}{dt}(\delta q_i) = \frac{d}{dt}\left(\frac{\partial L}{\partial \dot{q}_i}\delta q_i\right)$$

$$= \frac{d}{dt}(p_i \delta q_i) = \frac{d}{dt}(p_i \Delta q_i) - \frac{d}{dt}(p_i \dot{q}_i \Delta t) \qquad (6.33)$$

Combining (6.32) and (6.33) and allowing for the condition $\Delta q_i = 0$ at the end-points:

$$\Delta \int_1^2 L\,dt = \left[\left(L - \sum_i p_i \dot{q}_i\right)\Delta t\right]_1^2 = -\left[H\Delta t\right]_1^2 \qquad (6.34)$$

If consideration is restricted to systems for which $\frac{\partial H}{\partial t} = 0$ and to variations for which H remains constant then:

$$\left[H\Delta t\right]_1^2 = \Delta \int_1^2 H\,dt \qquad (6.35)$$

Substituting this result in (6.34) gives finally:

$$\Delta \int_1^2 \sum_i p_i \dot{q}_i\,dt = 0 \qquad (6.36)$$

This is the *principle of least action*. $W = \int \sum_i p_i \dot{q}_i\,dt$ is usually known as *Hamilton's characteristic function*. The type of variation involved is that represented by (6.30) with the additional restriction that the Hamiltonian is a constant of motion and takes the same value on the varied as on the unvaried path. The principle is not so directly useful in mechanics as Hamilton's principle, though it is invoked in Hamilton's development of the Hamilton–Jacobi method (see Chapter VII). It is of interest as being identical, in its essentials, with the original variational principle of Maupertuis and also with the earlier principle of Leibnitz in which he considered an integral of the *vis viva* of a system. Some confusion of terminology exists since some authors apply the term 'least action' to what has here been called Hamilton's principle.

For a detailed discussion of the many other variational principles which have been formulated at various times see, in particular, the work by Lanczos quoted in the bibliography.

CHAPTER VII

Transformation Theory

In building up the Lagrangian and Hamiltonian formulations considerable care was taken to make the *form* of all general relations the same for all co-ordinate systems. Any co-ordinate transformation represented by:

$$q_i' = q_i'(q_1, q_2, \ldots q_n, t) \tag{7.1}$$

produces equations identical in form with the original ones involving the q_i's. More technically it may be stated that 'the equations are covariant to point transformations'. It should be noted that the assumption is made that singular transformations are excluded from consideration.

In the Hamiltonian scheme momentum variables were introduced in addition to the position co-ordinates. These were defined through:

$$p_i = \frac{\partial L}{\partial \dot{q}_i} \tag{5.1}$$

At first sight these momentum variables appear to involve an entirely different concept from the position co-ordinates. Examination of Hamilton's canonical relations shows, however, that there is a certain similarity between the two sets of variables, thus:

$$\dot{q}_i = \frac{\partial H}{\partial p_i} \qquad \dot{p}_i = -\frac{\partial H}{\partial q_i} \tag{5.16}$$

and if the quantities q_i and p_i are relabelled $-p_i'$ and q_i' respectively equations (5.16) become:

$$\dot{q}_i' = \frac{\partial H}{\partial p_i} \qquad \dot{p}_i' = -\frac{\partial H}{\partial q_i'} \tag{7.2}$$

The equations (7.2) are indistinguishable in form from (5.16) and on a basis of (7.2) alone it would be natural to infer that the q_i' represent position co-ordinates and the p_i' momentum components.

In fact, according to the original definitions of these quantities, it is known that this is not so.

This results in an apparent paradox which can only be resolved by realizing that the q_i's and p_i's are to be treated on an equal basis. The original postulate of the Hamiltonian theory that the q_i's and p_i's are all to be regarded as independent variables is to be supplemented by the requirement that neither set is to be looked upon as more fundamental than the other.

An obvious corollary is to consider transformations of the type:

$$\left.\begin{array}{l} q_i' = q_i'(q_1, \ldots q_n, p_1, \ldots p_n, t) \\ p_i' = p_i'(q_1, \ldots q_n, p_1, \ldots p_n, t) \end{array}\right\} \tag{7.3}$$

which may be regarded as generalizations of (7.1).

Retaining the emphasis on general invariance it is required that the two sets of new variables shall be related through:

$$\dot{p}_i' = -\frac{\partial H'}{\partial q_i'} \qquad \dot{q}_i' = \frac{\partial H'}{\partial p_i'}$$

where H' is defined by:

$$H' = \sum_i p_i' \dot{q}_i' - L'$$

and L' is a function which when substituted in Hamilton's principle

$$\delta \int L' \, dt = 0$$

gives the correct equations of motion in terms of the new co-ordinates q_i'.

Transformations conforming to these conditions are given the description *contact* or *canonical*.* Transformations of the type (7.3) not fulfilling these requirements can be shown to exist but are of no practical interest.

In considering this more general type of transformation there has been a shift of viewpoint from that of Chapter III. There, in transforming to generalized co-ordinates, it was assumed that the new form of the function L was to be obtained from the old by direct

---

* Note that these two terms are not always regarded as synonymous, but here they will be used interchangeably.

substitution of the transformation relations. This constitutes a special case (termed a *point transformation*) of the more general type now being considered. Here there may be no direct relationship between the two forms of the Lagrangian.

It follows from the above that the modified form of Hamilton's principle holds in both the original and in the transformed system,

i.e.,

$$\left. \begin{aligned} \delta \int_{t_1}^{t_2} \left\{ \sum_i p_i \dot{q}_i - H \right\} dt &= 0 \\ \delta \int_{t_1}^{t_2} \left\{ \sum_i p_i' \dot{q}_i' - H' \right\} dt &= 0 \end{aligned} \right\} \tag{7.4}$$

An aspect of the δ variation process not previously considered is that the condition $\delta \int f \, dt = 0$ is satisfied, in general, by $f = \dfrac{dF}{dt}$ where F is any arbitrary function. This was not relevant to previous considerations where the integrand was a known function. Here, however, a different situation exists since equations (7.4) may be combined to give:

$$\delta \int_{t_1}^{t_2} \left\{ \left(\sum_i p_i \dot{q}_i - H \right) - \left(\sum_i p_i' \dot{q}_i' - H' \right) \right\} dt = 0 \tag{7.5}$$

in which the integrand is to a certain extent unknown. It follows that

$$\left(\sum_i p_i \dot{q}_i - H \right) - \left(\sum_i p_i' \dot{q}_i' - H' \right) = \frac{dF}{dt} \tag{7.6}$$

The first bracket of (7.6) is regarded as a function of the q_i, p_i and t, the second as a function of the q_i', p_i' and t. F is thus, in general, a function of the $(4n + 1)$ variables q_i, p_i, q_i', p_i' and t. However, the variables are subject to the transformation relations (7.3) and the dependence of F is reduced to $(2n + 1)$ independent variables, comprising t and any $2n$ of the q_i, p_i, q_i' and p_i'.

Consider the particular case:

$$F = F_1(q_1, \ldots q_n, q_1', \ldots q_n', t) \tag{7.7}$$

then:

$$\frac{dF_1}{dt} = \sum_i \frac{\partial F_1}{\partial q_i} \dot{q}_i + \sum_i \frac{\partial F_1}{\partial q_i'} \dot{q}_i' + \frac{\partial F_1}{\partial t} \tag{7.8}$$

Combining (7.6) with (7.8) gives:

$$\sum_i \left(p_i - \frac{\partial F_1}{\partial q_i} \right) dq_i - \sum_i \left(p_i' + \frac{\partial F_1}{\partial q_i'} \right) dq_i'$$

$$+ \left(H' - H - \frac{\partial F_1}{\partial t} \right) dt = 0 \qquad (7.9)$$

and since the q_i, q_i' and t may be regarded as independent variables:

$$p_i = \frac{\partial}{\partial q_i} F_1(q_i, q_i', t) \qquad (7.10)$$

$$p_i' = -\frac{\partial}{\partial q_i'} F_1(q_i, q_i', t) \qquad (7.11)$$

$$H' - H = \frac{\partial}{\partial t} F_1(q_i, q_i', t) \qquad (7.12)$$

(7.10) may (in principle) be solved giving:

$$q_i' = q_i'(q_i, p_i, t) \qquad (7.10')$$

substituting this in (7.11) then gives:

$$p_i' = p_i'(q_i, p_i, t) \qquad (7.11')$$

these are, of course, the transformation equations (7.3). It is thus seen that the transformation relations can be derived from a knowledge of the function F. On this account F is known as the *generating function* of the transformation.

It will be noted from (7.12) that the transformed Hamiltonian function is identical with the original unless the generating function contains t explicitly.

At first it seems that the details of a canonical transformation are fixed by what almost amounts to an arbitrary constant of integration. Closer inspection reveals that this is not a valid statement. In a situation in which the q_i, p_i, q_i' and p_i' are all predetermined there is no arbitrariness in F, it is a well defined function dependent upon the transformation relations; the fact that the latter can be derived from it should occasion no surprise. The misleading interpretation arises from it being easier to start from a given F and derive the transformation equations rather than to carry out the reverse process.

It was emphasized that the generating function can be expressed in terms of t and *any* $2n$ of the $4n$ variables q_i, p_i, q_i', p_i'. Any other case than that represented by (7.7) may be dealt with by effecting a Legendre transformation of this function: e.g., consider:

$$F_2 = F_1(q_1, \ldots q_n, q_1', \ldots q_n', t) + \sum_i p_i' q_i' \qquad (7.13)$$

From the general nature of Legendre transformations this would be expected to substitute the set (q_i, p_i', t) for (q_i, q_i', t) as independent variables. This implies:

$$F_2 = F_2(q_1, \ldots q_n, p_1', \ldots p_n', t) \qquad (7.13')$$

Considering the same transformation as before:

$$\left(\sum_i p_i \dot{q}_i - H \right) - \left(\sum_i p_i' \dot{q}_i' - H' \right) = \frac{dF_1}{dt} = \frac{d}{dt}\left(F_2 - \sum_i q_i' p_i' \right)$$

hence:

$$\sum_i \left(p_i - \frac{\partial F_2}{\partial q_i} \right) dq_i + \sum_i \left(q_i' - \frac{\partial F_2}{\partial p_i'} \right) dp_i'$$
$$+ \left(H' - H - \frac{\partial F_2}{\partial t} \right) dt = 0 \qquad (7.14)$$

from which:

$$p_i = \frac{\partial F_2}{\partial q_i} \qquad (7.15)$$

$$q_i' = \frac{\partial F_2}{\partial p_i'} \qquad (7.16)$$

$$H' - H = \frac{\partial F_2}{\partial t} \qquad (7.17)$$

Since $\dfrac{\partial F_1}{\partial t} = \dfrac{\partial F_2}{\partial t}$ * (7.17) is identical with (7.12) as would be expected from the fact that they refer to the same transformation. Also

* N.B. $\dfrac{\partial q_i}{\partial t}$ and other similar derivatives are zero since all the variables concerned are, by definition, independent.

(7.15) is identical with (7.10) since $\dfrac{\partial F_1}{\partial q_i} = \dfrac{\partial F_2}{\partial q_i}$. Equation (7.16) appears different from (7.11) but is really a rearrangement of it.

The relations relating to the other two main types of generating function are obtainable in similar fashion. They are summarized as follows:

$$F_3(p_i, q_i', t) = F_1(q_i, q_i', t) - \sum_i q_i p_i \qquad (7.18)$$

gives: $\qquad q_i = -\dfrac{\partial F_3}{\partial p_i} \qquad p_i' = -\dfrac{\partial F_3}{\partial q_i'} \qquad H' - H = \dfrac{\partial F_3}{\partial t} \qquad (7.19)$

and $\qquad F_4(p_i, p_i', t) = F_1(q_i, q_i', t) + \sum_i q_i' p_i' - \sum_i q_i p_i \quad (7.20)$

gives: $\qquad q_i = -\dfrac{\partial F_4}{\partial p_i} \qquad q_i' = \dfrac{\partial F_4}{\partial p_i'} \qquad H' - H = \dfrac{\partial F_4}{\partial t} \qquad (7.21)$

Examples of Canonical Transformations

(a) $F = \sum_i q_i p_i'$

This is obviously a special case of a generating function F_2, hence applying (7.15)–(7.17):

$$p_i = \frac{\partial F_2}{\partial q_i} = p_i' \qquad q_i' = \frac{\partial F_2}{\partial p_i'} = q_i \qquad H' = H + \frac{\partial F_2}{\partial t} = H$$

i.e., the transformation generated is the trivial one of the identity transformation.

$F = -\sum_i q_i p_i'$ generates the transformation $q_i' = -q_i$, $p_i' = -p_i$, $H' = H$. This illustrates the fact that space inversion constitutes a special case of a canonical transformation. The same is not true of simple time inversion.

(b) $F_1 = \sum_i q_i q_i'$

From (7.10)–(7.12):

$$p_i = q_i' \qquad p_i' = -q_i \qquad H = H'$$

which is the transformation considered in the opening remarks.

(c) $F_2 = \sum f_j(q_i)p_j'$ (f_j arbitrary)

$$\therefore p_i = \sum_j p_j' \frac{\partial f_j}{\partial q_i} \qquad q_i' = f_i(q_k) \qquad H' = H$$

This demonstrates the type of generating function required to generate a point transformation.

The above examples are merely illustrative of the process by which the transformation relations are derived given a specific form of generating function. It is not pretended that they in themselves are particularly useful. In fact, at this stage the question inevitably arises, 'what is the point of studying canonical transformations?' The aim of this book was to give an insight into a class of methods for formulating mechanical problems. If a problem has already been formulated in the form of Hamilton's canonical equations the only purpose of a contact transformation can be to put these equations into a more easily soluble form. That this can happen may be illustrated by considering a specific example.

Suppose that it is required to determine the motion of a particle for which a Hamiltonian function is given in the form:

$$H = \tfrac{1}{2}\left(\mu q^2 + \frac{p^2}{m}\right) \tag{7.22}$$

(this will, of course, be recognized as relating to a particle performing a linear S.H.M., but suppose this knowledge not to exist).

The postulation of (7.22) is equivalent to the following equations of motion:

$$\dot{q} = \frac{\partial H}{\partial p} = \frac{p}{m} \qquad \dot{p} = -\frac{\partial H}{\partial q} = -\mu q \tag{7.23}$$

In line with the statements made in Chapter V there is no obvious immediate solution to these equations. The simplest way to proceed would be to eliminate p giving:

$$m\ddot{q} = -\mu q \tag{7.24}$$

which is, in fact, Lagrange's equation for the system and is normally obtained without introducing the Hamiltonian function.

As an alternative procedure consider the contact transformation generated by the function:

$$F_1 = kq^2 \cot q' \tag{7.25}$$

for this transformation, from (7.10) to (7.12):

$$p = \frac{\partial F_1}{\partial q} = 2kq \cot q' \qquad H = H'$$

$$p' = - \frac{\partial F_1}{\partial q'} = kq^2 \operatorname{cosec}^2 q$$

hence:

$$p = \sqrt{4kp'} \,.\cos q' \qquad q = \sqrt{\frac{p'}{k}} \sin q' \tag{7.26}$$

$$\therefore H' = \frac{1}{2}\left(\mu q^2 + \frac{p^2}{m}\right) = \frac{1}{2}\left(\mu \frac{p'}{k} \sin^2 q' + \frac{4kp'}{m} \cos^2 q'\right)$$

$$= \frac{\mu p'}{2k}\left(\sin^2 q' + \frac{4k^2}{m\mu} \cos^2 q'\right) \tag{7.27}$$

if $k = \frac{1}{2}\sqrt{m\mu}$ this reduces to:

$$H' = \frac{\mu p'}{2k} = p'\sqrt{\frac{\mu}{m}} \tag{7.28}$$

which is of a particularly simple form. Since q' is an ignorable co-ordinate:

$$\dot{p}' = - \frac{\partial H'}{\partial q'} = 0$$

$$\therefore p' = \text{const.} = \alpha \text{ (say)} \tag{7.29}$$

also:

$$\dot{q}' = \frac{\partial H'}{\partial p'} = \sqrt{\frac{\mu}{m}} \ (= \text{const.})$$

$$\therefore q' = \sqrt{\frac{\mu}{m}} t + \beta \tag{7.30}$$

The desired expressions for p and q are now obtained by substituting the relations (7.29) and (7.30) in equations (7.26).

In this case the Hamilton's equations were rendered soluble by applying a contact transformation to effect a transformation to a new system in which the position co-ordinate was ignorable. Since the answer to the problem is so well known from other considerations it can only serve as an illustration of the general method, which is to make all the co-ordinates ignorable by a suitable choice of generating function. At this stage it is not apparent that anything more than pure guess-work is involved in the determination of this function. The development of a rational process will, however, form the subject matter of the next section.

As a by-product of the above example it is seen, using (7.26), (7.29) and (7.30) that:

$$J = \oint p \, dq = \int_0^{2\pi} 2p' \cos^2 q' \, dq' = 2\pi\alpha \ (= \text{const.}) \qquad (7.31)$$

J and q' are known as the action and angle variables. In the earlier forms of quantum mechanics the quantum behaviour of periodic systems was described by postulating that possible values of J must all be integral multiples of Planck's constant h.

The Hamilton–Jacobi Method

This offers a method for the explicit determination of a generating function from which can be derived a transformation which renders Hamilton's equations soluble. The type of transformation sought goes further than that considered previously in that all the new position and momentum co-ordinates are required to be constants.

Suppose that the desired transformation exists and is generated by the function $S = S(q_1, \ldots q_n, q_1', \ldots q_n', t)$, which must be a special case of the F_1 type of function considered in the last section. From the postulates above:

$$q_i' = \text{const.} = \alpha_i \qquad p_i' = \text{const.} = \beta_i \qquad (7.32)$$
$$\therefore \ S = S(q_1, \ldots q_n, \alpha_1, \ldots \alpha_n, t) \qquad (7.33)$$

Since the transformation is canonical:

$$\dot{q_i}' = \frac{\partial H'}{\partial p_i'} \qquad \dot{p_i}' = -\frac{\partial H'}{\partial q_i'}$$

and, allowing for (7.32):

$$\frac{\partial H'}{\partial p_i{}'} = 0 \qquad \frac{\partial H'}{\partial q_i{}'} = 0 \tag{7.34}$$

It is possible to impose the further condition that $\frac{\partial H'}{\partial t} = 0$. H' is then a constant which may be regarded as zero since, if the transformation S_0 gives rise to a transformed Hamiltonian $H_0 = \text{const.} = A$, then $S = S_0 - At$ gives $H' = 0$.

From (7.12): $\qquad H(q_i, p_i, t) + \dfrac{\partial S}{\partial t} = 0 \tag{7.35}$

also from (7.10): $\qquad p_i = \dfrac{\partial S}{\partial q_i} \tag{7.36}$

hence:

$$H\left(q_i, \frac{\partial S}{\partial q_i}, t\right) + \frac{\partial S}{\partial t} = 0 \tag{7.35'}$$

The equation (7.35') is a first order partial differential equation termed the *Hamilton–Jacobi equation*. It may be written down explicitly for any particular problem since the Hamiltonian will be a known function of the q_i, p_i and t. Its solution may present some difficulty but will be assumed possible in principle. We limit further discussion to the interpretation of the solution.

Since the Hamilton–Jacobi equation involves the $(n + 1)$ independent variables q_i and t its solution will contain $(n + 1)$ arbitrary constants. If S_0 is a possible solution it is obvious from the form of (7.35') that $S_1 = (S_0 + \text{const.})$ is likewise a solution. One of the $(n + 1)$ arbitrary constants is thus accounted for and may be ignored since only derivatives of S appear in the theory. The general solution may now be written:

$$S = S_0(q_1, \ldots q_n, c_1, \ldots c_n, t) \tag{7.37}$$

where the arbitrary constants are denoted by c_i.

However, it was originally given that:

$$S = S(q_1, \ldots q_n, q_1{}', \ldots q_n{}', t)$$

with $q_i{}' = \alpha_i = \text{const.}$ The c_i's must therefore be related to the α_i's and may in fact be identified with them. This identification requires that (7.32)–(7.34) be satisfied when c_i is substituted for α_i. The derivation has shown this to be the case except for the second of (7.32).

From (7.11) the generating function given by (7.37) transforms to momentum components given by:

$$p_i' = -\frac{\partial S_0}{\partial c_i} \tag{7.38}$$

$\therefore$ in view of (7.37):

$$\frac{dp_i'}{dt} = -\frac{d}{dt}\frac{\partial S_0}{\partial c_i} = -\left\{\sum_j \frac{\partial^2 S_0}{\partial q_j \partial c_i}\dot{q}_j + \frac{\partial^2 S_0}{\partial t \partial c_i}\right\} \tag{7.39}$$

also, since S_0 is a solution of (7.35'):

$$\frac{\partial^2 S_0}{\partial t \partial c_i} = \frac{\partial}{\partial c_i}\left(\frac{\partial S_0}{\partial t}\right) = -\frac{\partial}{\partial c_i}\left\{H\left(q_i, \frac{\partial S_0}{\partial q_i}, t\right)\right\} \tag{7.40}$$

from (7.37) the partial differentiation with respect to c_i is required to leave q_i and t constant, hence (7.40) becomes, using (7.36) and the canonical relation $\dot{q}_j = \dfrac{\partial H}{\partial p_j}$:

$$\frac{\partial^2 S_0}{\partial t \partial c_i} = -\sum_j \frac{\partial H}{\partial\left(\dfrac{\partial S_0}{\partial q_j}\right)}\frac{\partial}{\partial c_i}\left(\frac{\partial S_0}{\partial q_j}\right)$$

$$= -\sum_j \frac{\partial H}{\partial p_j}\frac{\partial^2 S_0}{\partial c_i \partial q_j} = -\sum_j \dot{q}_j\frac{\partial^2 S_0}{\partial c_i \partial q_j} \tag{7.41}$$

finally, combining (7.39) and (7.41) gives:

$$\frac{dp_i'}{dt} = 0 \qquad \text{i.e., } p_i' = \text{const.}$$

The identification of the c_i with the α_i thus satisfies all the original requirements of the transformation.

To determine the generating function S completely it is necessary to know the values of the constants $\alpha_i = c_i$. These may be found by substituting the given initial values of the problem (i.e., values of q_i and p_i at $t = t_0$) in equations (7.36) and solving these for the α_i. If, further, the values of the (constant) transformed momentum components $p_i' = \beta_i$ are required, these are found from (7.38).

This completes the problem as far as the details of the transformation are concerned, but the ultimate object will be the solution of the original Hamilton's equations of motion, i.e., the determination of the q_i and p_i as functions of time. To do this the α_i and β_i are substituted in the general form of equations (7.38) which are then solved for the q_i in the form $q_i = q_i(\alpha_j, \beta_j, t)$; finally the p_i are found by substituting the α_i and the q_i in (7.36) giving $p_i = p_i(\alpha_j, \beta_j, t)$.

Although it does not assist in the process of solution some further insight into the significance of the function S is obtained from the following considerations.

From (7.6):

$$\frac{dS}{dt} = \left(\sum_i p_i \dot{q}_i - H \right) - \left(\sum_i p_i' \dot{q}_i' - H' \right)$$

$$= \left(\sum_i p_i \dot{q}_i - H \right) = L$$

$$\therefore S = \int L \, dt \tag{7.42}$$

i.e., S is to be identified with Hamilton's principal function (hence the symbol employed). Unfortunately the integration cannot be performed until the q_i and p_i are known functions of the time, i.e., until the problem is solved. The identification is then, of course, of no practical use.

Example

In order to illustrate the Hamilton–Jacobi method the simple harmonic oscillator problem will again be considered:
For this problem

$$H = \tfrac{1}{2}\left(\frac{1}{m} p^2 + \mu q^2 \right) \tag{7.43}$$

and the corresponding Hamilton–Jacobi equation is:

$$\tfrac{1}{2}\left(\frac{1}{m}\left(\frac{\partial S}{\partial q} \right)^2 + \mu q^2 \right) + \frac{\partial S}{\partial t} = 0 \tag{7.44}$$

Whenever the Hamiltonian does not contain t explicitly the solution to the Hamilton–Jacobi equation takes the form:

$$S(q, \alpha, t) = S'(q, \alpha) - c(\alpha)t$$

According to the discussion of the general case the constant $c(\alpha)$ may be identified with α itself. Thus:

$$S(q, \alpha, t) = S'(q, \alpha) - \alpha t \tag{7.45}$$

and (7.44) becomes:

$$\tfrac{1}{2}\left(\frac{1}{m}\left(\frac{\partial S'}{\partial q}\right)^2 + \mu q^2\right) - \alpha = 0$$

i.e.,

$$\frac{\partial S'}{\partial q}(= p) = \sqrt{m\mu}\left(\frac{2\alpha}{\mu} - q^2\right)^{\frac{1}{2}} \tag{7.46}$$

$$\therefore\ S = S' - \alpha t = \sqrt{m\mu}\int\left(\frac{2\alpha}{\mu} - q^2\right)^{\frac{1}{2}} dq + D(\alpha) - \alpha t \tag{7.47}$$

where D is a constant of integration which, as was pointed out previously, can be ignored without loss of generality.

In this case it is unnecessary to obtain the explicit form of S by performing the integration. From (7.47):

$$\beta = -\frac{\partial S}{\partial \alpha} = t - \sqrt{m\mu}\int\frac{1}{\mu}\left(\frac{2\alpha}{\mu} - q^2\right)^{-\frac{1}{2}} dq$$

$$= \left\{t - \sqrt{\frac{m}{\mu}}\cos^{-1}\left(q\sqrt{\frac{\mu}{2\alpha}}\right)\right\} \tag{7.48}$$

Taking the initial conditions as $q = 0$, $p = \sqrt{2mE_0}$ at $t = 0$ and substituting in (7.46) gives:

$$\alpha = E_0 \tag{7.49}$$

i.e., the transformed generalized co-ordinate is to be identified with the total energy.

Also, from (7.48):

$$\beta = -\frac{\pi}{2}\sqrt{\frac{m}{\mu}} \tag{7.50}$$

The explicit form of q is now obtained from (7.48) as:

$$q = \sqrt{\frac{2E_0}{\mu}}\cos\left(t\sqrt{\frac{\mu}{m}} + \frac{\pi}{2}\right) \tag{7.51}$$

and then from (7.46):

$$p = \sqrt{2mE_0} \cdot \sin\left(t\sqrt{\frac{\mu}{m}} + \frac{\pi}{2}\right) \quad (7.52)$$

these, of course, represent the familiar solution to the problem.

When there is more than one pair of canonical variables and t does not appear explicitly in H it is always true that the solution of the H–J equation is of the form:

$$S(q_i, \alpha_i, t) = S'(q_i, \alpha_i) - \alpha t \quad (7.53)$$

where α is identifiable with the total energy in all cases of interest.* The full determination of S is, however, possible only when the variables q_i are separable in the H–J equation.

Infinitesimal Contact Transformations

When discussing examples of canonical transformations it was shown that the function $F_2 = \sum_i q_i p_i'$ generates the identity transformation. If ε is an infinitesimal parameter independent of q_i and p_i it follows that an infinitesimal change in the variables will be generated by:

$$F = \sum_i q_i p_i' + \varepsilon G(q_i, p_i') \quad (7.54)$$

where G is an arbitrary function. From (7.15) and (7.16) the new variables are given by:

$$q_i' = \frac{\partial F}{\partial p_i'} = q_i + \varepsilon\frac{\partial G}{\partial p_i'} \qquad p_i = \frac{\partial F}{\partial q_i} = p_i' + \varepsilon\frac{\partial G}{\partial q_i} \quad (7.55)$$

i.e.,
$$\delta q_i = \varepsilon\frac{\partial G}{\partial p_i'} \qquad \delta p_i = -\varepsilon\frac{\partial G}{\partial q_i} \quad (7.56)$$

and since $(p_i' - p_i)$ is infinitesimal it suffices to replace p_i' by p_i in the function G giving:

$$\delta q_i = \varepsilon\frac{\partial}{\partial p_i}G(q_i, p_i) \qquad \delta p_i = -\varepsilon\frac{\partial}{\partial q_i}G(q_i, p_i) \quad (7.56')$$

* See the section on 'Geometrical and Wave Mechanics'.

Although it was originally the function F which was referred to as the generating function, in the case of infinitesimal contact transformations it is customary to transfer the description to the function G. Equations (7.56′) thus give the infinitesimal changes in the conjugate variables which are generated by an arbitrary function G.

As a particular case consider:

$$\varepsilon = dt \qquad G = H \qquad (7.57)$$

then, using the canonical relations:

$$\delta q_i = dt\frac{\partial H}{\partial p_i} = dt.\dot{q}_i \qquad \delta p_i = -\,dt\frac{\partial H}{\partial q_i} = dt.\dot{p}_i \qquad (7.58)$$

i.e., the changes in the conjugate variables produced by using the Hamiltonian as a generating function are those actually occurring in the system in the course of its motion. The changes in a finite period from t_0 to t may be considered built up from a succession of such infinitesimal changes all generated by H. The motion of the system can thus be regarded as the continuous unfolding of a transformation generated by the Hamiltonian of the system.

Geometrical and Wave Mechanics

In the light of the last section it is now possible to place a somewhat different interpretation on the procedure of the Hamilton–Jacobi method. Previously it was considered as a means for solving problems by transforming to a new canonical description in which all the variables are constants of motion. This is the interpretation due to Jacobi. An alternative view, first put forward by Hamilton, is to regard S as a function which transforms the position co-ordinates from their initial values q_i' at $t = 0$ to their values q_i at time t. It thus summarizes the development of the system in time.

By definition $\qquad L = \sum_i p_i\dot{q}_i - H$

$$\therefore S = \int L\,dt = \int \sum_i p_i\dot{q}_i\,dt - \int H\,dt = W - \int H\,dt \qquad (7.59)$$

From initial assumption, $S = S(q_i, q_i', t)$. Hence confining attention to systems for which $\dfrac{\partial H}{\partial t} = 0$ and $H = E$ and omitting the $q_i{}'$ since they are now to be regarded as the (constant) initial values:

$$S(q_i, t) = W - Et \tag{7.60}$$

The principle of least action (6.36) asserts that $\Delta W = 0$. From the significance of the Δ variation given in Chapter VI it is clear that W can depend only upon the position co-ordinates of the end-points of the path.

$$\therefore \; S(q_i, t) = W(q_i) - Et \tag{7.60'}$$

as stated previously in (7.53).

It is possible to develop a general interpretation of (7.60') in terms of wave motion. We shall content ourselves by doing so for a simple special case which shows more clearly the considerations involved.

Consider a single particle system in which the forces are conservative and the co-ordinate system is a cartesian one. In general the Hamilton–Jacobi equation is:

$$H\left(q_i, \frac{\partial S}{\partial q_i}, t\right) + \frac{\partial S}{\partial t} = 0 \tag{7.35'}$$

for our particular case we have from (7.36) and (7.60') that:

$$p_i = \frac{\partial S}{\partial q_i} = \frac{\partial W}{\partial q_i} \quad \text{i.e.,} \quad \mathbf{p} = \nabla W \tag{7.61a}$$

also:

$$\frac{\partial S}{\partial t} = -E \tag{7.61b}$$

hence the Hamilton–Jacobi equation is:

$$\frac{1}{2m}(\nabla W)^2 + V = E \tag{7.62}$$

i.e.,

$$|\nabla W| = \sqrt{2m(E - V)} \tag{7.62'}$$

It has already been suggested that the motion of a system may be represented by a continuous curve in configuration space. In the present case this curve will be the actual trajectory of the particle in ordinary space. $W = $ constant represents a set of surfaces in this space and (7.61a) implies that the trajectory of the particle is everywhere normal to such surfaces. This is suggestive of the relation between wave surfaces and rays in optics. Suppose that the particle motion is in fact associated with some form of wave motion in this way. If the wave behaviour is represented by a wave function ψ obeying an equation similar to the scalar wave equation in optics, then:

$$\nabla^2\psi - \frac{\mu^2}{v_0^2}\frac{\partial^2\psi}{\partial t^2} = 0 \qquad (7.63)$$

where allowance has been made for a variation of wave velocity from point to point by the incorporation of a 'refractive index', μ, which

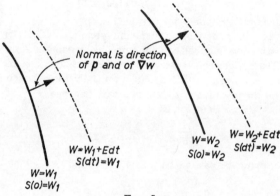

Normal is direction of $\mathbf{p}$ and of ∇w

$W = W_1 + Edt$
$S(dt) = W_1$

$W = W_1$
$S(o) = W_1$

$W = W_2$
$S(o) = W_2$

$W = W_2 + Edt$
$S(dt) = W_2$

FIG. 5

is a continuous function of position. Confining attention to a single frequency ω:

$$\nabla^2\psi - \frac{\mu^2\omega^2}{v_0^2}\psi = 0 \qquad (7.63')$$

The general solution to this equation is:

$$\psi = \psi_0(q_i)e^{i[k_0 f(q_i) - \omega t]} \qquad (7.64)$$

where $k_0 = 2\pi/\lambda_0 = \omega/v_0$, and ψ_0 is real.

With a restriction to cases where the variation of refractive index is small in a single wavelength this may be shown to require the condition that:

$$(\nabla f)^2 = \mu^2 \qquad (7.65)$$

Surfaces of constant phase are given by:

$$k_0 f(q_i) - \omega t = \phi(q_i, t) = \text{const.} \tag{7.66}$$

This is analogous with (7.60′) and it is thus possible to identify surfaces of constant S (which may also be plotted in configuration space and will coincide instantaneously with various W surfaces as shown in Fig. 5) as wave surfaces of constant phase. From this identification:

$$S = a\phi, \qquad W = ak_0 f, \qquad E = a\omega \tag{7.67}$$

where a is some constant. (7.65), (7.67) and (7.62′) now give as a value for the 'refractive index':

$$\mu = |\nabla f| = \frac{1}{ak_0} |\nabla W| = \frac{v_0}{\omega a} \sqrt{2m(E - V)} \tag{7.68}$$

the wave equation (7.63′) thus becomes:

$$\nabla^2 \psi - \frac{2m(E - V)}{a^2} \psi = 0 \tag{7.69}$$

Putting $a = h/2\pi$, this is recognizable as Schrödinger's wave equation for the single particle in a conservative field. Schrödinger's wave mechanics is thus seen to bear the same relation to ordinary particle mechanics as does physical optics to geometrical (or ray) optics.[*] On this account particle mechanics is often referred to as geometrical mechanics.

Fermat's Principle

The analogy developed in the preceding section may be pursued in reverse. For the particle case, substitution of (7.61a) and (7.68) in the principle of least action (6.36) gives:

$$\Delta \int \mu \, ds = 0 \tag{7.70}$$

Applied to the optical case this result is *Fermat's principle of least optical path*.

[*] For an account of recent work in this field see J. L. Synge, *Geometrical Mechanics and de Broglie Waves* (C.U.P., 1954).

CHAPTER VIII

Poisson Brackets

Definition

Let $F = F(q_i, p_i, t)$ be any dynamical variable of a system represented by the conjugate variables q_i, p_i. Then:

$$\dot{F} \equiv \frac{dF}{dt} = \sum_i \frac{\partial F}{\partial q_i} \dot{q}_i + \sum_i \frac{\partial F}{\partial p_i} \dot{p}_i + \frac{\partial F}{\partial t} \tag{8.1}$$

From Hamilton's canonical equations (5.16) this becomes:

$$\dot{F} = \sum_i \left(\frac{\partial F}{\partial q_i} \frac{\partial H}{\partial p_i} - \frac{\partial F}{\partial p_i} \frac{\partial H}{\partial q_i} \right) + \frac{\partial F}{\partial t} \tag{8.2}$$

The quantity $\sum_i \left(\dfrac{\partial F}{\partial q_i} \dfrac{\partial H}{\partial p_i} - \dfrac{\partial F}{\partial p_i} \dfrac{\partial H}{\partial q_i} \right)$ turns out to be a very significant one in the formal development of mechanics and is called the Poisson bracket of F and H. In general, the Poisson bracket of any two dynamical variables X and Y is defined as:

$$\left[X, Y \right]_{q,\,p} = \sum_i \left(\frac{\partial X}{\partial q_i} \frac{\partial Y}{\partial p_i} - \frac{\partial X}{\partial p_i} \frac{\partial Y}{\partial q_i} \right) \tag{8.3}$$

The concept does not assist materially in the complete solution of the equations of motion of a system, but is of use in discussing the constants of motion, as will be seen. It leads to a formalism which, when re-interpreted according to a simple recipe, forms a convenient way of introducing quantum rules in the Heisenberg development of quantum mechanics.

The following identities follow immediately from the definition:

$$\left.\begin{array}{c} \left[X, Y\right] = -\left[Y, X\right] \\ \left[X, X\right] = 0 \\ \left[X, Y + Z\right] = \left[X, Y\right] + \left[X, Z\right] \\ \left[X, YZ\right] = Y\left[X, Z\right] + \left[X, Y\right]Z \end{array}\right\} \qquad (8.4)$$

also:

$$\left.\begin{array}{c} \left[q_i, q_j\right]_{q, p} = 0 = \left[p_i, p_j\right]_{q, p} \\ \left[q_i, p_j\right]_{q, p} = \delta_{ij} \end{array}\right\} \qquad (8.5)$$

where δ_{ij} is the usual delta symbol with the property:

$$\begin{array}{cc} \delta_{ij} = 0 & i \neq j \\ = 1 & i = j \end{array}$$

The quantities (8.5) are known as the *fundamental* or *basic* Poisson brackets.

Invariance with Respect to Canonical Transformations

A highly important property of the brackets is their invariance with respect to a canonical transformation. By this statement is meant:

$$\left[X, Y\right]_{q, p} = \left[X, Y\right]_{q', p'}$$

where it is understood that X and Y have the same *value* but not necessarily the same *form* in terms of the two sets of co-ordinates.

The proof of the above statement can be verified using ideas developed in the last chapter. It was shown there that a canonical transformation could be generated from a function $F_1 = F_1(q_i, q_i', t)$, in which case the following relations would hold:

$$p_i = \frac{\partial F_1}{\partial q_i} \qquad (7.10)$$

$$p_i' = -\frac{\partial F_1}{\partial q_i'} \qquad (7.11)$$

from these it follows that:

$$\frac{\partial p_i}{\partial q_j'} = \frac{\partial^2 F_1}{\partial q_j' \partial q_i} = -\frac{\partial p_j'}{\partial q_i} \tag{8.6}$$

and in similar fashion using the other types of generating function F_2, F_3, F_4:

$$\frac{\partial q_i}{\partial q_j'} = \frac{\partial p_j'}{\partial p_i} \qquad \frac{\partial q_i}{\partial p_j'} = -\frac{\partial q_j'}{\partial p_i} \qquad \frac{\partial p_i}{\partial p_j'} = \frac{\partial q_j'}{\partial q_i} \tag{8.6'}$$

Applying these results to the fundamental Poisson brackets:

$$\begin{aligned}
\left[q_i', p_j'\right]_{q,\,p} &\equiv \sum_k \left(\frac{\partial q_i'}{\partial q_k}\frac{\partial p_j'}{\partial p_k} - \frac{\partial q_i'}{\partial p_k}\frac{\partial p_j'}{\partial q_k}\right) \\
&= \sum_k \frac{\partial q_i'}{\partial q_k}\left(\frac{\partial p_j'}{\partial q_k} + \frac{\partial p_i'}{\partial q_k}\frac{\partial q_k}{\partial p_k}\right) \\
&= \sum_k \frac{\partial q_i'}{\partial q_k}\frac{\partial p_j'}{\partial p_k} = \sum_k \frac{\partial q_i'}{\partial q_k}\frac{\partial q_k}{\partial q_j'} = \delta_{ij} \\
&= \left[q_i', p_j'\right]_{q',\,p'}
\end{aligned}$$

and similarly:
$$\left.\begin{aligned}
\left[q_i', q_j'\right]_{q,\,p} &= 0 = \left[q_i', q_j'\right]_{q',\,p'} \\
\left[p_i', p_j'\right]_{q,\,p} &= 0 = \left[p_i', p_j'\right]_{q',\,p'}
\end{aligned}\right\} \tag{8.7}$$

The assertion has thus been proved for the fundamental brackets. Consider now the general case:

$$\begin{aligned}
\left[X, Y\right]_{q',\,p'} &\equiv \sum_k \left(\frac{\partial X}{\partial q_k'}\frac{\partial Y}{\partial p_k'} - \frac{\partial X}{\partial p_k'}\frac{\partial Y}{\partial q_k'}\right) \\
&= \sum_{j,\,k} \left\{\frac{\partial X}{\partial q_k'}\left(\frac{\partial Y}{\partial q_j}\frac{\partial q_j}{\partial p_k'} + \frac{\partial Y}{\partial p_j}\frac{\partial p_j}{\partial p_k'}\right) \right. \\
&\qquad\qquad \left. - \frac{\partial X}{\partial p_k'}\left(\frac{\partial Y}{\partial q_j}\frac{\partial q_j}{\partial q_k'} + \frac{\partial Y}{\partial p_j}\frac{\partial p_j}{\partial q_k'}\right)\right\} \\
&= \sum_j \left\{\frac{\partial Y}{\partial q_j}\left[X, q_j\right]_{q',\,p'} + \frac{\partial Y}{\partial p_j}\left[X, p_j\right]_{q',\,p'}\right\} \tag{8.8}
\end{aligned}$$

$$\left[q_j, X\right]_{q', p'}$$ may be considered as a special case of (8.8) by making the substitutions q_j for X and X for Y.

$$\therefore \left[X, q_j\right]_{q', p'} = -\left[q_j, X\right]_{q', p'}$$

$$= -\sum_k \left\{\frac{\partial X}{\partial q_k}\left[q_j, q_k\right]_{q', p'} + \frac{\partial X}{\partial p_k}\left[q_j, p_k\right]_{q', p'}\right\}$$

$$= -\sum_k \frac{\partial X}{\partial p_k}\delta_{jk} = -\frac{\partial X}{\partial p_j} \quad \text{[using (8.7)]} \quad (8.9)$$

similarly:
$$\left[X, p_j\right]_{q', p'} = \frac{\partial X}{\partial q_j} \quad (8.10)$$

substituting (8.9) and (8.10) in (8.8):

$$\left[X, Y\right]_{q', p'} = \sum_j \left\{-\frac{\partial Y}{\partial q_j}\frac{\partial X}{\partial p_j} + \frac{\partial Y}{\partial p_j}\frac{\partial X}{\partial q_j}\right\} \equiv \left[X, Y\right]_{q, p} \quad (8.11)$$

thus verifying the result in the general case. Since the value of a Poisson bracket is independent of the set of conjugate variables with respect to which it is evaluated the subscript on the bracket is unnecessary and will now be omitted.

Angular Momentum

Angular momentum components have been identified with generalized momentum components in particular cases. In general the momentum conjugate to any angular co-ordinate can be identified in this way in a simple mechanical system where, for instance, electromagnetic effects are not present. It is of interest to investigate the Poisson bracket of two components of angular momentum. For simplicity consider a particle referred to a cartesian co-ordinate system, the angular momentum components are then given by:

$$l_1 = x_2p_3 - x_3p_2 \qquad l_2 = x_3p_1 - x_1p_3 \qquad l_3 = x_1p_2 - x_2p_1 \quad (8.12)$$

where $p_1 = m\dot{x}_1$, etc.

Evaluating the Poisson bracket of l_1 and l_2 gives:

$$\left[l_1, l_2\right] = (p_2x_1 - p_1x_2) = l_3$$

Similar results are obtained for other combinations and they may be summarized in:

$$\left[l_i, l_j \right] = \sum_k \varepsilon_{ijk} l_k \tag{8.13}$$

where $\varepsilon_{ijk} = \quad 1$ if (i, j, k) is an even permutation of $(1, 2, 3)$
$\quad\quad\quad = -1$,,   ,,   ,, ,, odd   ,,   ,, $(1, 2, 3)$
$\quad\quad\quad = \quad 0$ otherwise.

The implication of (8.13) is that no two components of angular momentum can simultaneously act as conjugate momenta since all conjugate variables must obey the laws relating to the fundamental brackets given in (8.7). Any one angular momentum component can, of course, be chosen as a generalized momentum co-ordinate but not more than one in any particular system of reference.

Consider now $\left[l_i, l^2 \right]$, where l^2 is the square of the total angular momentum. Using the identities (8.4) and the result (8.13):

$$\left[l_i, l^2 \right] = \left[l_i, \sum_j l_j^2 \right] = \sum_j \left[l_i, l_j^2 \right] = \sum_j \left\{ 2l_j \left[l_i, l_j \right] \right\}$$

$$= \sum_{j, k} 2l_j \varepsilon_{ijk} l_k \equiv 0 \tag{8.14}$$

i.e., l^2 and any one component of l can simultaneously be regarded as conjugate momenta. The results (8.13) and (8.14) are highly important in the extension of the formalism to quantum mechanics.

Other results with similar significance are:

$$\left[x_i, l_j \right] = \sum_k \varepsilon_{ijk} x_k \qquad \left[p_i, l_j \right] = \sum_k \varepsilon_{ijk} p_k \tag{8.15}$$

where the p's in this instance still denote linear cartesian momentum components.

Constants of Motion

It has already been emphasized that for some purposes the solution of a problem may be considered achieved by identifying the constants of motion. Rewriting (8.2) in Poisson bracket notation shows that the time variation of any dynamical variable F is given by:

$$\dot{F} = \left[F, H \right] + \frac{\partial F}{\partial t} \tag{8.2'}$$

This shows that if the variable does not contain the time explicitly it is sufficient for its Poisson bracket with H to vanish in order that it be a constant of motion. This result is independent of whether H itself is a constant of motion and provides a useful means for identifying constants of motion.

Special cases of (8.2') are:

$$\dot{q}_i = \left[q_i, H \right] \qquad \dot{p}_i = \left[p_i, H \right] \tag{8.16}$$

these are identical with Hamilton's canonical equations and may be referred to as the equations of motion in Poisson bracket form.

Another special case is:

$$\frac{dH}{dt} = \left[H, H \right] + \frac{\partial H}{\partial t} = \frac{\partial H}{\partial t} \tag{8.17}$$

This relation has also appeared previously.

Jacobi's Identity

Consider the expression:

$$\left[X, \left[Y, Z \right] \right] - \left[Y, \left[X, Z \right] \right]$$

$$= \left[X, \sum_i \left(\frac{\partial Y}{\partial q_i} \frac{\partial Z}{\partial p_i} - \frac{\partial Y}{\partial p_i} \frac{\partial Z}{\partial q_i} \right) \right] - \left[Y, \sum_i \left(\frac{\partial X}{\partial q_i} \frac{\partial Z}{\partial p_i} - \frac{\partial X}{\partial p_i} \frac{\partial Z}{\partial q_i} \right) \right]$$

using the identities (8.4) and regrouping gives:

$$\sum_i \left\{ - \frac{\partial Z}{\partial q_i} \left(\left[\frac{\partial X}{\partial p_i}, Y \right] + \left[X, \frac{\partial Y}{\partial p_i} \right] \right) + \frac{\partial Z}{\partial p_i} \left(\left[\frac{\partial X}{\partial q_i}, Y \right] + \left[X, \frac{\partial Y}{\partial q_i} \right] \right) \right\}$$

$$+ \sum_i \left\{ \frac{\partial Y}{\partial q_i} \left[X, \frac{\partial Z}{\partial p_i} \right] - \frac{\partial Y}{\partial p_i} \left[X, \frac{\partial Z}{\partial q_i} \right] - \frac{\partial X}{\partial q_i} \left[Y, \frac{\partial Z}{\partial p_i} \right] + \frac{\partial X}{\partial p_i} \left[Y, \frac{\partial Z}{\partial q_i} \right] \right\}$$

7

Using the identity:

$$\frac{\partial}{\partial x}\Big[X,\,Y\Big] \equiv \Big[\frac{\partial X}{\partial x},\,Y\Big] + \Big[X,\,\frac{\partial Y}{\partial x}\Big] \tag{8.18}$$

the first expression reduces to:

$$\sum_i \left\{ -\frac{\partial Z}{\partial q_i}\frac{\partial}{\partial p_i}\Big[X,\,Y\Big] + \frac{\partial Z}{\partial p_i}\frac{\partial}{\partial q_i}\Big[X,\,Y\Big] \right\} = -\Big[Z,\,\Big[X,\,Y\Big]\Big]$$

the second expression may be shown to vanish. Hence:

$$\Big[X,\,\Big[Y,\,Z\Big]\Big] - \Big[Y,\,\Big[X,\,Z\Big]\Big] = -\Big[Z,\,\Big[X,\,Y\Big]\Big]$$

which may be written in symmetrical form:

$$\Big[X,\,\Big[Y,\,Z\Big]\Big] + \Big[Y,\,\Big[Z,\,X\Big]\Big] + \Big[Z,\,\Big[X,\,Y\Big]\Big] = 0 \tag{8.19}$$

The result is known as Jacobi's identity. It may be applied in the following way. Let $Z = H$, then:

$$\Big[X,\,\Big[Y,\,H\Big]\Big] + \Big[Y,\,\Big[H,\,X\Big]\Big] + \Big[H,\,\Big[X,\,Y\Big]\Big] = 0 \tag{8.20}$$

if now X and Y are both constants of motion then:

$$\Big[Y,\,H\Big] = 0 \qquad \Big[X,\,H\Big] = 0 \tag{8.21}$$

hence:
$$\Big[H,\,\Big[X,\,Y\Big]\Big] = 0 \tag{8.22}$$

i.e., the dynamical variable $\Big[X,\,Y\Big]$ is also a constant of motion. The usefulness of this result lies in the possibility of constructing new constants of motion from known ones. It will not, however, always be the case that the new constants are other than trivialities (e.g., using p_i and p_j = const. merely leads to 0 = const.).

Poisson Brackets and Commutators

In quantum mechanics dynamical variables are represented by operators which do not obey the commutation rules of ordinary algebra. It is not possible to define Poisson brackets for these operators, but the universal nature and general usefulness of the brackets in classical mechanics suggests that there might be analogous quantities associated with the operators.

Equations (8.4) can be held to represent basic properties of Poisson brackets. Assume that these also represent properties of quantities associated with the corresponding quantum mechanical operators. (Note that this is not a necessary assumption and that, in fact, it is not true of all types of operator.) For any three operators X, Y, Z we have, on this assumption:

$$\left[X, YZ\right] = Y\left[X, Z\right] + \left[X, Y\right]Z$$

and
$$\left[XY, Z\right] = X\left[Y, Z\right] + \left[X, Z\right]Y$$

where for initial convenience we employ the same symbol to denote this unknown quantum analogue of the Poisson bracket. Care has been taken to preserve the order of the operators in view of their non-commutability.

It follows that, for any four operators W, X, Y, Z:

$$\left[WX, YZ\right] = W\left[X, YZ\right] + \left[W, YZ\right]X$$
$$= W\left[X, Y\right]Z + WY\left[X, Z\right] + Y\left[W, Z\right]X + \left[W, Y\right]ZX$$

also, expanding in a different order:

$$\left[WX, YZ\right] = \left[WX, Y\right]Z + Y\left[WX, Z\right]$$
$$= W\left[X, Y\right]Z + \left[W, Y\right]XZ + YW\left[X, Z\right] + Y\left[W, Z\right]X$$

Combining these results gives:

$$(WY - YW)\left[X, Z\right] = \left[W, Y\right](XZ - ZX) \qquad (8.23)$$

The four operators were assumed arbitrary. It follows that the identity (8.23) can only be satisfied if, for any two operators A, B:

$$(AB - BA) = \alpha\left[A, B\right] \qquad (8.24)$$

where α is some constant, i.e., the quantum analogue of the Poisson bracket is identified as a multiple of the commutator of the two operators concerned. Assuming that the operators corresponding to the conjugate variables q_i, p_i play the same fundamental role as the classical variables, then:

$$(q_i p_j - p_j q_i) = \alpha \delta_{ij} \tag{8.25}$$

It is by further postulating that $\alpha = \dfrac{ih}{2\pi}$ that the quantum rules are introduced in the Heisenberg formulation of quantum mechanics.

Infinitesimal Contact Transformations

These were considered briefly in the previous chapter. There it was shown that a generating function:

$$F = \sum_i q_i p_i{}' + \varepsilon G(q_i, p_i{}') \tag{7.54}$$

where G is arbitrary, produces changes in the conjugate variables given by:

$$\delta q_i = \varepsilon \frac{\partial G}{\partial p_i} \qquad \delta p_i = - \varepsilon \frac{\partial G}{\partial q_i} \tag{7.56'}$$

where the infinitesimal nature of the transformation has been taken advantage of in substituting p_i for $p_i{}'$ in G.

Corresponding to these changes in the q_i and p_i there will be a change in the value of any dynamical variable $X = X(q_i, p_i, t)$ given by:

$$\delta X = \sum_i \frac{\partial X}{\partial q_i} \delta q_i + \sum_i \frac{\partial X}{\partial p_i} \delta p_i \tag{8.26}$$

hence from (7.56'):

$$\delta X = \varepsilon \sum_i \left(\frac{\partial X}{\partial q_i} \frac{\partial G}{\partial p_i} - \frac{\partial X}{\partial p_i} \frac{\partial G}{\partial q_i} \right) = \varepsilon \Big[X, G \Big] \tag{8.27}$$

If $G = H$ and $\varepsilon = dt$ (8.27) becomes:

$$\delta X = dt \Big[X, H \Big] \tag{8.28}$$

In the case where $X = q_i$ this gives:

$$\delta q_i = dt \Big[q_i, H \Big] = \dot{q}_i \, dt$$

which can be interpreted in the sense that the infinitesimal transformation generated by H gives rise to the actual motion of the system. The same applies when X is identified with any of the other conjugate variables p_i. Caution must however be exercised in interpreting the general case for, from (8.2′):

$$\left[X, H\right] = \frac{dX}{dt} - \frac{\partial X}{\partial t}$$

and (8.28) becomes:
$$\delta X = \left(\frac{dX}{dt} - \frac{\partial X}{\partial t}\right)dt \qquad (8.28')$$

from which it is seen that the change δX generated by the transformation is not the actual change in X which will occur in the course of the motion unless $\frac{\partial X}{\partial t} = 0$.

The general interpretation of (8.28) is thus that, as far as it concerns any dynamical variables not explicitly dependent upon t, the motion of a system is equivalent to a sequence of infinitesimal contact transformations generated by the Hamiltonian.

Another aspect of (8.27) is seen by substituting H for X:

$$\delta H = \varepsilon\left[H, G\right] \qquad (8.29)$$

From this it is seen that a constant of motion $\left(\text{for which } \left[H, G\right] = 0\right)$ generates an infinitesimal transformation with respect to which H is invariant. In the special case $G = p_x = $ (const. of motion), the transformation concerned is an infinitesimal translation along the x-axis. This may be seen using (7.56′) which give:

$$\delta x = \varepsilon \qquad \delta y = \delta z = 0 \qquad \delta p_x = \delta p_y = \delta p_z = 0$$

Hence the Hamiltonian is invariant to translations for which the corresponding linear momentum component is unchanged. This is merely another way of expressing the fact that an ignorable co-ordinate implies a correspondingly constant momentum component. A similar result follows from a consideration of infinitesimal rotations generated by functions G identified with angular momentum components. These last considerations have brought us back to the considerations of Chapter V where the connection between symmetry and constants of motion was discussed. The introduction of an argument based on Poisson brackets has led to a broadening of the field to include all constants of motion rather than just constant momenta as in the previous discussion. It has now been shown that the Hamiltonian is invariant (and hence the system symmetric) with respect to any infinitesimal transformation generated by a constant of motion. The reverse statement is also true and makes it possible to detect constants of motion by inspecting the Hamiltonian for any symmetries it contains.

CHAPTER IX

Continuous Systems

The Lagrangian and Hamiltonian methods have been developed so far to deal with discrete systems having a finite number of degrees of freedom. It is the purpose of the present chapter to extend these to include continuous systems in which the number of degrees of freedom is infinitely many. This extension is not a difficult matter once a suitable Lagrangian function has been identified; there is, however, an element of surprise in the form of the parameters upon which the various functions depend.

Lagrangian Formulation

An insight into the modifications required can be obtained by developing the considerations of previous sections. A special case of a continuous system would be a continuous one-dimensional elastic solid and this may be treated as the limiting case of a linear chain of interacting particles. For convenience the interaction will be considered to be supplied by springs connecting each pair of neighbouring particles.

Let the particles be all of mass m, distance a apart and connected by springs with force constants k. As before, denote the displacement of the ith particle by η_i. Its equation of motion is then from the definition of force constant:

$$m\ddot{\eta}_i = k\{(\eta_{i+1} - \eta_i) - (\eta_i - \eta_{i-1})\} \qquad (9.1)$$

The system is a conservative one and the forces may be derived from a scalar potential function:*

$$V = \sum_i \tfrac{1}{2}k(\eta_{i+1} - \eta_i)^2 \qquad (9.2)$$

---

* A difficulty of principle arises here in ensuring that all particles obey the same equation of motion. If the chain is of finite length it is necessary

The kinetic energy of the system is:

$$T = \sum_i \tfrac{1}{2} m \dot{\eta}_i^2 \tag{9.3}$$

hence the Lagrangian function:

$$L = T - V = \sum_i \{\tfrac{1}{2} m \dot{\eta}_i^2 - \tfrac{1}{2} k(\eta_{i+1} - \eta_i)^2\} = \sum a L_i \tag{9.4}$$

A simple check shows that the Lagrange's equations obtained by differentiating in the usual way are identical with the equations of motion (9.1).

The discrete system may now be made to pass over into a continuous one with the following correspondences:

$$a \rightarrow dx \qquad \sum (\quad)a \rightarrow \int (\quad)dx$$

$$\frac{m}{a} \rightarrow \rho \text{ (density)} \qquad ka \rightarrow E \text{ (Young's modulus)} \tag{9.5}$$

$$\frac{\eta_{i+1} - \eta_i}{a} \rightarrow \frac{d\eta}{dx}$$

hence: $$L \rightarrow \int \mathscr{L} \, dx \tag{9.6}$$

where:* $$\mathscr{L} = \tfrac{1}{2}\left\{\rho\dot{\eta}^2 - E\left(\frac{d\eta}{dx}\right)^2\right\} \tag{9.7}$$

It would thus appear that the Lagrangian function which relates to the whole system under consideration can be considered as an

---

to assume the existence of additional boundary forces. If it is infinitely long, then functions such as V diverge unless periodic boundary conditions are assumed, making it necessary only to consider that length occupying a single period. In what follows the problem will be assumed resolved in one of these ways.

* Note at this point the introduction of the use of total differential symbols in situations where it will be more usual to find partial differential ones. It is fundamental to the treatment that x and t are independent variables and there is no call to denote this fact in any special way.

Subsequently it will be necessary to differentiate functions which depend on x and t through their explicit dependence on η and its derivatives. The more general convention to be used then is that:

$$\frac{d}{dt} \equiv \frac{\partial}{\partial t} + \frac{d\eta}{dt}\frac{\partial}{\partial \eta} + \frac{d\eta_{,x}}{dt}\frac{\partial}{\partial \eta_{,x}} + \frac{d\dot{\eta}}{dt}\frac{\partial}{\partial \dot{\eta}}$$

integral of another (density) function. The function $\mathscr{L}$ is termed the *Lagrangian density* of the system.

If Hamilton's principle is held to apply we can write:

$$\delta \int L \, dt = 0 \qquad (9.8)$$

In view of (9.6) this may be re-expressed as:

$$\delta \int \int \mathscr{L} \, dx \, dt = 0 \qquad (9.9)$$

There is, of course, no guarantee that Hamilton's principle will in fact apply and the only test will be a comparison of the equations of motion derived from the assumed principle with their form derived by an alternative method.

A Hamilton's principle of the form (9.9) has not previously been encountered, but it suggests that x and t should now be treated on an equal footing. The variation of the integral is therefore carried out by varying the path of integration in such a way as to keep the end points (η_1, x_1, t_1 and η_2, x_2, t_2) of the path fixed and to vary the value of η at all other points but keeping x and t fixed. The necessary and sufficient conditions for such a variation to give a stationary value to the integral were quoted in Chapter VI (equations 6.9); translated into present terms they are:

$$\left(\frac{\partial \mathscr{L}}{\partial \eta} - \frac{d}{dt} \frac{\partial \mathscr{L}}{\partial \dot{\eta}} - \frac{d}{dx} \frac{\partial \mathscr{L}}{\partial \eta_{,\,x}} \right) = 0 \qquad (9.10)$$

where $\eta_{,\,x} \equiv \dfrac{d\eta}{dx}$.

Applying (9.10) to the case under consideration gives:

$$\frac{d}{dt}(\rho\dot{\eta}) - E\frac{d^2\eta}{dx^2} = 0$$

i.e.,
$$\rho\ddot{\eta} = E\frac{d^2\eta}{dx^2} \qquad (9.11)$$

with a similar expression for $\dfrac{d}{dx}$. The generalizations to three space coordinates and to more than one field variable η are obvious.

This usage provides a consistent notation and avoids the introduction of new symbols.

This is recognizable as the correct equation for the propagation of waves in a one-dimensional elastic solid and is usually obtained as a limiting form of (9.1). The assumption that Hamilton's principle may be applied to continuous systems has thus led to consistency in the particular case under consideration.

In all cases which can be tested there is similar agreement and it is concluded that the mechanics of a continuous system can be summarized in the mathematical statement:

$$\delta \int L \, dt = \delta \iiint \mathscr{L} \, dt \, dx_1 \, dx_2 \, dx_3 \equiv \delta \iint \mathscr{L} \, dV \, dt = 0 \quad (9.12)$$

implying the possibility of finding a density function

$$\mathscr{L} \equiv \mathscr{L}\left(\eta^{(r)}, \frac{d\eta^{(r)}}{dt}, \frac{d\eta^{(r)}}{dx_j}, x_j, t\right)$$

from which may be deduced the correct equations of motion of the system. Allowance has here been made for extension to three (cartesian) space co-ordinates and to n variables $\eta^{(r)}$. The nature of these variables (usually termed *field variables*) is not, in general, confined to displacements as in the above elastic problem. The formalism is found, for instance, to be suitable for the description of an electromagnetic field in which case there are no less than four field variables consisting of the scalar potential ϕ and the three components of the vector potential A. This aspect will be developed in greater detail in Chapter XI after the relativistic considerations of Chapter X.

The conditions for (9.12) to hold are n equations of the form:

$$\frac{\partial \mathscr{L}}{\partial \eta^{(r)}} - \frac{d}{dt}\frac{\partial \mathscr{L}}{\partial \dot{\eta}^{(r)}} - \sum_j \frac{d}{dx_j}\frac{\partial \mathscr{L}}{\partial \eta_{,j}^{(r)}} = 0 \qquad (9.13)$$

with $\eta_{,j}^{(r)} = \dfrac{d\eta^{(r)}}{dx_j}$.

The quantity $\left(\dfrac{\partial \mathscr{L}}{\partial \eta} - \displaystyle\sum_j \dfrac{d}{dx_j}\dfrac{\partial \mathscr{L}}{\partial \eta_{,j}}\right)$ is often written $\dfrac{\delta L}{\delta \eta}$ and is

referred to as the *functional derivative* of L. Apart from brevity there appears to be no particular gain from this symbolism.

Since second derivatives are postulated not to occur in the Lagrangian density function, $\dfrac{\partial \mathscr{L}}{\partial \dot{\eta}_{,j}} \equiv 0$, hence:

$$\frac{\partial \mathscr{L}}{\partial \dot{\eta}} = \frac{\partial \mathscr{L}}{\partial \dot{\eta}} - \sum_j \frac{d}{dx_j} \frac{\partial \mathscr{L}}{\partial \dot{\eta}_{,j}} \equiv \frac{\delta L}{\delta \dot{\eta}}$$

(9.13) may thus be written:

$$\frac{\delta L}{\delta \eta^{(r)}} = \frac{d}{dt} \frac{\delta L}{\delta \dot{\eta}^{(r)}} \tag{9.13'}$$

Admittedly this is a concise relation analogous in form to the corresponding result for the discrete case, but it conveys no further information than (9.13).

The net result of these considerations is that, in the case of continuous systems, the field variables $\eta^{(r)}$ take over the role played by the space co-ordinates in the discrete case. The space co-ordinates still appear in the formulation, but they join the time variable t as independent parameters. Again anticipating the content of the following chapters, it may be remarked that this symmetrical treatment of time and space co-ordinates has prepared the way for the incorporation of relativistic considerations.

Hamiltonian Formulation

This again follows much the same lines as for the discrete case and, apart from the use of the density functions and the substitution of functional for ordinary partial derivatives, the results are similar in form.

In our *discrete case* the momentum conjugate to the variable η_i is defined by:

$$p_i = \frac{\partial L}{\partial \dot{\eta}_i} \tag{9.14}$$

In the transition to the continuum:

$$L = \sum_i a L_i \longrightarrow \int \mathscr{L} \, dx \tag{9.6}$$

$$\therefore p_i \equiv \frac{\partial L}{\partial \dot{\eta}_i} = a \frac{\partial L_i}{\partial \dot{\eta}_i} \longrightarrow \frac{\partial \mathscr{L}}{\partial \dot{\eta}} dx \tag{9.15}$$

i.e., a definition of conjugate momentum for the continuous case made in direct analogy with that for the discrete case leads to an infinitesimal quantity. The alternative is to define a *momentum density*:

$$\pi = \frac{\partial \mathscr{L}}{\partial \dot{\eta}} \tag{9.16}$$

which is, of course, finite.

In general there will be n field variables $\eta^{(r)}$ and n corresponding *conjugate variables* (or *canonical momentum densities*) $\pi^{(r)} = \dfrac{\partial \mathscr{L}}{\partial \dot{\eta}^{(r)}}$. In line with the previous development a *Hamiltonian density* is defined through:

$$\mathscr{H} = \sum_r \pi^{(r)} \dot{\eta}^{(r)} - \mathscr{L} \tag{9.17}$$

with an integral Hamiltonian given by:

$$H = \int \mathscr{H} \, dV \tag{9.18}$$

It is to be noted that:

$$\mathscr{H} = \mathscr{H}\left(\eta^{(r)}, \frac{d\eta^{(r)}}{dx_j}, \pi^{(r)}, x_j, t\right) \tag{9.19}$$

since the $\dot{\eta}^{(r)}$ dependence is supposed eliminated. In the Hamiltonian theory the $\eta^{(r)}$ and $\pi^{(r)}$ are all regarded as independently variable functions of the x_i and t. This is in contrast to the Lagrangian description where only the $\eta^{(r)}$ are assumed to be independently variable. The situation is, of course, analogous to that in particle mechanics. A common aim will be the determination of the generalized field variables as explicit functions of the space and time co-ordinates in the same way that, in particle mechanics, the generalized co-ordinates are determined as functions of t.

In the specific example considered $L = T - V$ and $H = T + V$ = total energy. However, it is difficult to formulate general conditions for the Hamiltonian to be identified with the total energy of the system. As far as the equations of motion derived from it are concerned, any Lagrangian may be multiplied by an arbitrary constant without altering its properties; hence the first requirement in the energy identification is to fix this multiplying constant so that one or more terms of the Lagrangian are recognizably energy quantities. (In the discrete case this was effected automatically by ensuring that at least one term in the equations of motion was recognizable as a genuine force component in the Newtonian sense.)

The second condition is that the co-ordinate axes should be at rest with respect to the frame of reference. This is not a serious restriction since moving axes are rarely encountered in connection with continuous systems. It is, however, worth stating explicitly in view of the example quoted in the discussion of the same point in Chapter V.

Provided the above restrictions hold and the system is non-dissipative it is usually assumed that the Hamiltonian and the total energy are identical. Cases other than those involving purely conservative systems should, strictly speaking, be investigated individually, as in the case of the particle in the electromagnetic field. The matter of this identification is highly important since the chief point of the Hamiltonian function is as a measure of the total energy which can be calculated without identifying force components as in the Newtonian scheme.

Hamilton's Canonical Equations of Motion

Consider now the total differential of $\mathscr{H}$:

$$
\left.
\begin{aligned}
d\mathscr{H} = \sum_r \left\{ \frac{\partial \mathscr{H}}{\partial \eta^{(r)}} d\eta^{(r)} + \sum_j \frac{\partial \mathscr{H}}{\partial \eta_{,\,j}{}^{(r)}} d\eta_{,\,j}{}^{(r)} + \frac{\partial \mathscr{H}}{\partial \pi^{(r)}} d\pi^{(r)} \right\} \\
+ \sum_j \frac{\partial \mathscr{H}}{\partial x_j} dx_j + \frac{\partial \mathscr{H}}{\partial t} dt
\end{aligned}
\right\} \qquad (9.20)
$$

also from (9.17):

$$
\begin{aligned}
d\mathscr{H} &= \sum_r \left\{ \dot{\eta}^{(r)}\, d\pi^{(r)} + \pi^{(r)}\, d\dot{\eta}^{(r)} \right\} - d\mathscr{L} \\
&= \sum_r \left\{ \dot{\eta}^{(r)}\, d\pi^{(r)} + \frac{\partial \mathscr{L}}{\partial \dot{\eta}^{(r)}} d\dot{\eta}^{(r)} \right\} - \sum_r \left\{ \frac{\partial \mathscr{L}}{\partial \eta^{(r)}} d\eta^{(r)} + \sum_j \frac{\partial \mathscr{L}}{\partial \eta_{,\,j}{}^{(r)}} d\eta_{,\,j}{}^{(r)} \right. \\
&\qquad \left. + \frac{\partial \mathscr{L}}{\partial \dot{\eta}^{(r)}} d\dot{\eta}^{(r)} \right\} - \sum \frac{\partial \mathscr{L}}{\partial x_j} dx_j - \frac{\partial \mathscr{L}}{\partial t} dt
\end{aligned}
$$

$$\left.\begin{aligned} &= \sum_r \left\{ \dot{\eta}^{(r)} \, d\pi^{(r)} - \frac{\partial \mathscr{L}}{\partial \eta^{(r)}} d\eta^{(r)} - \sum_j \frac{\partial \mathscr{L}}{\partial \eta,_j^{(r)}} d\eta,_j^{(r)} \right\} \\ &\qquad - \sum_j \frac{\partial \mathscr{L}}{\partial x_j} dx_j - \frac{\partial \mathscr{L}}{\partial t} dt \end{aligned}\right\} \quad (9.21)$$

Equating coefficients in (9.20) and (9.21):

$$\left.\begin{aligned} \frac{\partial \mathscr{H}}{\partial \eta^{(r)}} = -\frac{\partial \mathscr{L}}{\partial \eta^{(r)}} \quad \frac{\partial \mathscr{H}}{\partial \eta,_j^{(r)}} = -\frac{\partial \mathscr{L}}{\partial \eta,_j^{(r)}} \quad \frac{\partial \mathscr{H}}{\partial \pi^{(r)}} = \dot{\eta}^{(r)} \\ \frac{\partial \mathscr{H}}{\partial t} = -\frac{\partial \mathscr{L}}{\partial t} \quad \frac{\partial \mathscr{H}}{\partial x_j} = -\frac{\partial \mathscr{L}}{\partial x_j} \end{aligned}\right\} \quad (9.22)$$

the first three of these relations could be termed the Hamiltonian equations of motion but are not yet in the accepted form. From (9.13), (9.16) and (9.22):

$$\left.\begin{aligned} \dot{\pi}^{(r)} &\equiv \frac{d}{dt}\left(\frac{\partial \mathscr{L}}{\partial \dot{\eta}^{(r)}}\right) = \frac{\partial \mathscr{L}}{\partial \eta^{(r)}} - \sum_j \frac{d}{dx_j} \frac{\partial \mathscr{L}}{\partial \eta,_j^{(r)}} \\ &= -\frac{\partial \mathscr{H}}{\partial \eta^{(r)}} + \sum_j \frac{d}{dx_j} \frac{\partial \mathscr{H}}{\partial \eta,_j^{(r)}} \equiv -\frac{\delta H}{\delta \eta^{(r)}} \end{aligned}\right\} \quad (9.23)$$

also: $\qquad \dot{\eta}^{(r)} = \frac{\partial \mathscr{H}}{\partial \pi^{(r)}} = \frac{\partial \mathscr{H}}{\partial \pi^{(r)}} - \sum_j \frac{d}{dx_j} \frac{\partial \mathscr{H}}{\partial \pi,_j^{(r)}} \equiv \frac{\delta H}{\delta \pi^{(r)}}$

since derivatives of $\pi^{(r)}$ do not appear in $\mathscr{H}$.

The relations analogous to the previous canonical equations are thus:

$$\frac{\delta H}{\delta \pi^{(r)}} = \dot{\eta}^{(r)} \qquad \frac{\delta H}{\delta \eta^{(r)}} = -\dot{\pi}^{(r)} \quad (9.23')$$

It may also be noted that:

$$\frac{\partial \mathscr{H}}{\partial x_i} = -\frac{\partial \mathscr{L}}{\partial x_i} \qquad \frac{\partial \mathscr{H}}{\partial t} = -\frac{\partial \mathscr{L}}{\partial t} \quad (9.24)$$

though explicit dependence upon the time and space co-ordinates rarely occurs.

Density Conservation Laws

It is of interest to investigate the time variation of the Hamiltonian density. From (9.21) and (9.22) remembering that t and the x_j are independent variables:

$$\frac{d\mathscr{H}}{dt} = \sum_r \left\{ \left(\dot{\pi}^{(r)} - \frac{\partial \mathscr{L}}{\partial \eta^{(r)}} \right) \dot{\eta}^{(r)} - \sum_j \frac{\partial \mathscr{L}}{\partial \eta,_j^{(r)}} \dot{\eta},^{(r)} \right\} + \frac{\partial \mathscr{H}}{\partial t} \qquad (9.21')$$

Also from Lagrange's equations of motion:

$$\frac{\partial \mathscr{L}}{\partial \eta^{(r)}} - \frac{d}{dt} \frac{\partial \mathscr{L}}{\partial \dot{\eta}^{(r)}} - \sum_j \frac{d}{dx_j} \frac{\partial \mathscr{L}}{\partial \eta,_j^{(r)}} = 0 \qquad (9.13)$$

i.e.,

$$\left(\dot{\pi}^{(r)} - \frac{\partial \mathscr{L}}{\partial \eta^{(r)}} \right) = - \sum_j \frac{d}{dx_j} \frac{\partial \mathscr{L}}{\partial \eta,_j^{(r)}}$$

Substituting this result in (9.21'):

$$\frac{d\mathscr{H}}{dt} = - \sum_j \sum_r \left\{ \frac{d}{dx_j} \left(\frac{\partial \mathscr{L}}{\partial \eta,_j^{(r)}} \right) \dot{\eta}^{(r)} + \frac{\partial \mathscr{L}}{\partial \eta,_j^{(r)}} \dot{\eta},_j^{(r)} \right\} + \frac{\partial \mathscr{H}}{\partial t}$$

$$= - \sum_j \frac{dS_j}{dx_j} + \frac{\partial \mathscr{H}}{\partial t} \qquad (9.25)$$

where:
$$S_j = \sum_r \frac{\partial \mathscr{L}}{\partial \eta,_j^{(r)}} \dot{\eta}^{(r)} \qquad (9.26)$$

In cases of interest $\dfrac{\partial \mathscr{H}}{\partial t} = 0$ and (9.25) takes the form:

$$\nabla . \mathbf{S} + \frac{d\mathscr{H}}{dt} = 0 \qquad (9.25')$$

Identifying $\mathscr{H}$ as the total energy density, this equation may be interpreted as an energy continuity equation, or *conservation law*, with $\mathbf{S}$ as a vector representing the flow of energy. $\mathbf{S}$ is not unique since (9.25') holds for any vector $\mathbf{S}' = \mathbf{S} + \nabla \wedge \mathbf{X}$, where $\mathbf{X}$ is an arbitrary vector and $\mathbf{S}$ is given by (9.26). It is, however, usual to simplify considerations by considering the special case where $\nabla \wedge \mathbf{X} = 0$.

The form of the density quantities $\mathscr{H}$ and $\mathbf{S}$ suggests an investigation of analogous density quantities defined by:

$$\mathscr{G}_j = - \sum_r \frac{\partial \mathscr{L}}{\partial \dot{\eta}^{(r)}} \eta,_j^{(r)} \qquad (9.27)$$

The only obvious property of such quantities is that, irrespective of the nature of the $\eta^{(r)}$, they have the dimensions of ordinary linear momentum. They may, in fact, be identified with particular quantities of this kind as may be seen by referring once more to the simple case of the one-dimensional elastic solid.

Consider a slab inside the solid of thickness Δx. The planes defining this slab are supposed fixed with respect to the axes of reference and it follows that the displacement, represented by the field variable η, will result in a movement of the medium through the planes. Consider unit area of the planes. The mass of the medium which has

$$\eta \qquad \left(\eta + \frac{d\eta}{dx}\Delta x\right)$$
$$\rightarrow \qquad \rightarrow$$
$$x \qquad x + \Delta x$$

moved *into* the slab through the face at x is $\rho\eta$ and that which has moved *out* through the second face at $(x + \Delta x)$ is $\rho\left(\eta + \frac{d\eta}{dx}\Delta x\right)$.

The net mass which has moved *into* the slab is thus $-\rho\dfrac{d\eta}{dx}\Delta x$.

This results in a net surplus of momentum in the slab of amount $-\rho\dfrac{d\eta}{dx}\dot{\eta}\Delta x$ per unit area, i.e., the density of the excess momentum in the slab is $-\rho\dot{\eta}\dfrac{d\eta}{dx}$. This quantity is not to be confused with the actual *momentum density of the medium* $\rho\dot{\eta}$. It represents a new *differential* quantity which may be termed the *wave momentum density* since it will only be non-zero for a wave motion in which $\dfrac{d\eta}{dx} \neq 0$.

It will be recalled that a Lagrangian density function suitable for describing the properties of the one dimensional elastic solid was:

$$\mathscr{L} = \tfrac{1}{2}\left(\rho\dot{\eta}^2 - E\left(\frac{d\eta}{dx}\right)^2\right) \qquad (9.7)$$

It is readily seen that, in this one dimensional case, the single quantity defined by (9.27) would be $\mathscr{G} = -\rho\dot{\eta}\dfrac{d\eta}{dx}$, which is identical with the quantity just investigated and described as the wave momentum density. A similar identification can be made in other cases where the field variables represent displacements of a material medium. In the general case (9.27) is held to define the wave (or field) momentum density.

The question now arises whether there are any conservation laws relating to the wave (or field) momentum. Consider the time derivative of $\mathscr{G}_j$:

$$\frac{d\mathscr{G}_j}{dt} = -\frac{d}{dt}\left\{\sum_r \frac{\partial \mathscr{L}}{\partial \dot{\eta}^{(r)}}\eta,_j^{(r)}\right\}$$

$$= -\sum_r \left\{\frac{d}{dt}\left(\frac{\partial \mathscr{L}}{\partial \dot{\eta}^{(r)}}\right)\eta,_j^{(r)} + \frac{\partial \mathscr{L}}{\partial \dot{\eta}^{(r)}}\frac{d}{dt}\left(\eta,_j^{(r)}\right)\right\}$$

$$= -\sum_r \left\{\frac{\partial \mathscr{L}}{\partial \eta^{(r)}}\frac{d\eta^{(r)}}{dx_j} - \sum_i \frac{d}{dx_i}\left(\frac{\partial \mathscr{L}}{\partial \eta,_i^{(r)}}\right)\frac{d\eta^{(r)}}{dx_j} + \frac{\partial \mathscr{L}}{\partial \dot{\eta}^{(r)}}\frac{d\dot{\eta}^{(r)}}{dx_j}\right\}$$

[using (9.13)]

$$= -\frac{d\mathscr{L}}{dx_j} + \frac{\partial \mathscr{L}}{\partial x_j} + \sum_r \sum_i \frac{d}{dx_i}\left(\frac{\partial \mathscr{L}}{\partial \eta,_i^{(r)}}\frac{d\eta^{(r)}}{dx_j}\right)$$

$$= \sum_i \frac{d}{dx_i}\sum_r \left\{\frac{\partial \mathscr{L}}{\partial \eta,_i^{(r)}}\frac{d\eta^{(r)}}{dx_j} - \delta_{ij}\mathscr{L}\right\} + \frac{\partial \mathscr{L}}{\partial x_j} \tag{9.28}$$

If x_j does not appear explicitly in $\mathscr{L}$, then $\dfrac{\partial \mathscr{L}}{\partial x_j} = 0$. With this not very severe restriction the components of the wave momentum density are seen to obey conservation laws which may be written:

$$\frac{d\mathscr{G}_j}{dt} = \sum_i \frac{dT_{ij}}{dx_i} \tag{9.29}$$

The quantities $T_{ij} = \sum_r \left\{\dfrac{\partial \mathscr{L}}{\partial \eta,_i^{(r)}}\dfrac{d\eta^{(r)}}{dx_j} - \delta_{ij}\mathscr{L}\right\}$ are components of a second order tensor identifiable as a stress tensor. Note, however, that the system of stresses concerned is that acting through planes fixed with respect to the axes of reference and not moving with the medium. It would also be possible to regard these quantities as components of vectors representing momentum flow, in analogy with the energy flow vector **S**.

Integral Conservation Laws and Poisson Brackets

From (9.25) and again assuming that $\mathscr{L}$ and $\mathscr{H}$ do not depend explicitly on the time we have:

$$\frac{dH}{dt} = \frac{d}{dt}\int \mathscr{H}\,dV = \int \frac{d\mathscr{H}}{dt}dV = -\int \nabla.\mathbf{S}\,dV = -\int \mathbf{S}.d\boldsymbol{\sigma} \quad (9.30)$$

where the last integral is over the bounding surface of the region of integration. This integral vanishes if the system is of finite size and is contained within the region of integration. It may also be shown to vanish in the case of an infinite system with periodic boundary conditions. In each of these cases, therefore, the total energy is a constant of motion.

A similar consideration holds for the total wave or field momentum $\mathbf{G} = \int \mathscr{G}\,dV$ since, from (9.28):

$$\frac{dG_j}{dt} = \int \frac{d\mathscr{G}_j}{dt}dV = \int \sum_i \frac{dT_{ij}}{dx_i}dV + \int \frac{\partial\mathscr{L}}{\partial x_j}dV \quad (9.31)$$

In all cases of interest $\dfrac{\partial\mathscr{L}}{\partial x_j}\left(= -\dfrac{\partial\mathscr{H}}{\partial x_j}\right) = 0$; also the divergence

integral vanishes under the same conditions as before. Hence **G** is a constant of motion under essentially the same restrictive conditions as is H.

A wave angular momentum density may be defined through:

$$\mathscr{M}_{ij} = x_i\mathscr{G}_j - x_j\mathscr{G}_i$$

This definition is consistent with previous usage of the term angular momentum. The time variation of the total wave angular momentum components is given by:

$$\frac{dM_{ij}}{dt} = \frac{d}{dt}\int \mathscr{M}_{ij}\,dV = \int \left(x_i\frac{d\mathscr{G}_j}{dt} - x_j\frac{d\mathscr{G}_i}{dt}\right)dV$$

$$= \int \sum_k \left(x_i \frac{dT_{kj}}{dx_k} - x_j \frac{dT_{ki}}{dx_k} \right) dV \quad \text{[using (9.29)]}$$

$$= \int \sum_k \frac{d}{dx_k}(x_i T_{kj} - x_j T_{ki}) \, dV - \int (T_{ij} - T_{ji}) \, dV \quad (9.32)$$

The divergence term vanishes under the same assumptions as before, leaving:

$$\frac{dM_{ij}}{dt} = \int (T_{ji} - T_{ij}) dV \tag{9.32'}$$

It is usual to postulate that the total angular momentum of the system is a constant of motion. From (9.32') this is seen to imply that the stress tensor T_{ij} is a symmetrical one. The requirement is usually found to be satisfied automatically with mechanical systems. In the exceptional cases it is possible to effect a symmetrization by adding suitable terms to the Lagrangian density. This possibility is connected with the degree of arbitrariness already noted in connection with the energy flow vector S. The matter will be reconsidered in more general terms in Chapter XI.

As in particle mechanics the identification of constants of motion plays an important role and it is of interest to determine the time variation of a general integral dynamical variable given by:

$$Y = \int \mathcal{Y} \, dV \tag{9.33}$$

where:
$$\mathcal{Y} = \mathcal{Y}(\eta^{(r)}, \eta_{,\,i}^{(r)}, \pi^{(r)}, x_i, t) \tag{9.34}$$

$$\frac{dY}{dt} = \int \frac{d\mathcal{Y}}{dt} dV = \int \left\{ \sum_r \left(\frac{\partial \mathcal{Y}}{\partial \eta^{(r)}} \dot{\eta}^{(r)} + \sum_i \frac{\partial \mathcal{Y}}{\partial \eta_{,\,i}^{(r)}} \frac{d}{dt} \eta_{,\,i}^{(r)} \right. \right.$$
$$\left. \left. + \frac{\partial \mathcal{Y}}{\partial \pi^{(r)}} \dot{\pi}^{(r)} \right) + \frac{\partial \mathcal{Y}}{\partial t} \right\} dV$$

$$\left. \begin{array}{r} = \int \left\{ \sum_r \left(\frac{\partial \mathcal{Y}}{\partial \eta^{(r)}} - \sum_i \frac{d}{dx_i} \frac{\partial \mathcal{Y}}{\partial \eta_{,\,i}^{(r)}} \right) \dot{\eta}^{(r)} + \frac{\partial \mathcal{Y}}{\partial \pi^{(r)}} \dot{\pi}^{(r)} \right\} dV \\ + \int \frac{\partial \mathcal{Y}}{\partial t} dV \end{array} \right\} \tag{9.35}$$

Here, the second term has been integrated by parts and it is assumed, as before, that the $\dot{\eta}^{(r)}$ vanish over the boundaries or that there are periodic boundary conditions.

Invoking the functional derivative notation and the canonical equations (9.23), (9.35) becomes:

$$\frac{dY}{dt} = \int \sum_r \left(\frac{\delta Y}{\delta \eta^{(r)}} \frac{\delta H}{\delta \pi^{(r)}} - \frac{\delta Y}{\delta \pi^{(r)}} \frac{\delta H}{\delta \eta^{(r)}} \right) dV + \int \frac{\partial \mathscr{Y}}{\partial t} dV \quad (9.36)$$

In this form the result bears some resemblance to the analogous result for the discrete case. This may be emphasized by making an extended Poisson bracket definition:

$$\left[X, Y \right] \equiv \int \sum_r \left(\frac{\delta X}{\delta \eta^{(r)}} \frac{\delta Y}{\delta \pi^{(r)}} - \frac{\delta X}{\delta \pi^{(r)}} \frac{\delta Y}{\delta \eta^{(r)}} \right) dV \quad (9.37)$$

(9.36) then becomes:

$$\frac{dY}{dt} = \left[Y, H \right] + \frac{\partial Y}{\partial t} \quad (9.38)$$

which is now formally identical with the result for the discrete case.

This new definition provides no further information. It can be shown, however, that the quantities concerned are precisely analogous to the Poisson brackets considered in Chapter VIII, and it may be inferred that they have the same sort of universal significance. The definition is confined to integral quantities and, in particular, there are no quantities corresponding to the fundamental brackets encountered in the discrete case.

Transition to Quantum Mechanics

The absence of fundamental Poisson brackets for a continuous system implies that the transition to quantum mechanics cannot be carried out exactly as for a discrete system. A method for resolving this difficulty is indicated by referring once more to our elementary example. In the case of the discrete elastically bound particles the canonical momenta are given by:

$$p_i = \frac{\partial L}{\partial \dot{\eta}_i} = a \frac{\partial L_i}{\partial \dot{\eta}_i} \quad (9.15)$$

Here the original Poisson bracket definition applies and we have

$$\left[\eta_i, p_j\right] = \delta_{ij} \tag{9.39}$$

i.e.,

$$\sum_j \left[\eta_i, p_j\right] = \sum_j \delta_{ij} = 1 \tag{9.39'}$$

In the transition to quantum mechanics this leads, as discussed in Chapter VIII, to a corresponding relation for the commutator of the *operators* η_i and p_j:

$$\sum_j (\eta_i p_j - p_j \eta_i) = \alpha \quad (= ih/2\pi) \tag{9.40}$$

Assuming that, in the transition to a continuum, there are operator correspondences identical with those for the classical dynamical variables, i.e., assuming that:

$$\eta_i \longrightarrow \eta(x) \qquad p_j \longrightarrow \pi(x')dx' \tag{9.41}$$

then (9.40) becomes:

$$\int \{\eta(x)\pi(x') - \pi(x')\eta(x)\}dx' = \alpha \tag{9.42}$$

this may be interpreted as:

$$\{\eta(x)\pi(x') - \pi(x')\eta(x)\} = \alpha \cdot \delta(x - x') \tag{9.43}$$

where $\delta(x - x')$ is the Dirac delta function defined by:

$$\left. \begin{aligned} \delta(x - x') &= 0 \qquad x \neq x' \\ \int f(x)\delta(x - x')dx &= f(x') \end{aligned} \right\} \tag{9.44}$$

and it is supposed that the region of integration includes x'.

The result may again be generalized to n field operators $\eta^{(r)}$ together with their associated conjugate momentum operators $\pi^{(r)}$, also to three space dimensions, giving:

$$\{\eta^{(r)}(\mathbf{r})\pi^{(s)}(\mathbf{r}') - \pi^{(s)}(\mathbf{r}')\eta^{(r)}(\mathbf{r})\} = \alpha\delta_{rs}\delta(\mathbf{r} - \mathbf{r}') \tag{9.45}$$

with

$$\delta(\mathbf{r} - \mathbf{r}') = \delta(x_1 - x_1')\delta(x_2 - x_2')\delta(x_3 - x_3')$$

The stipulation that $\alpha = ih/2\pi$ supplies the normal means of postulating the quantum behaviour of a continuous system.

Applications

In this chapter there has been developed a somewhat formidable-looking apparatus to extend the Lagrangian and Hamiltonian formalisms to continuous systems. It has been seen that many of the features of the discrete case can be generalized without difficulty. The considerations could be extended to show the existence of the previous correspondence between constants of motion and the symmetry properties of a system.*

The formalism was developed with specific reference to a simple type of material system. It is found readily applicable to a variety of more complicated material systems.† Thus the general equations for the propagation of elastic waves in isotropic solids may be deduced from a Lagrangian density function of the form:

$$\mathscr{L} = \tfrac{1}{2} \sum_i \left\{ (\lambda + \mu) \sum_j \frac{d\eta_i}{dx_j} \frac{d\eta_j}{dx_i} + \mu \sum_j \left(\frac{d\eta_i}{dx_j} \right)^2 - \rho \left(\frac{d\eta_i}{dt} \right)^2 \right\}$$

where η is the vector displacement at any point, μ is the shear modulus of the medium and $(\lambda + 2\mu/3)$ is its bulk modulus. The field variables are not necessarily displacements. The irrotational motion of a compressible non-viscous fluid in the so-called acoustic approximation is deducible from the density function

$$\mathscr{L} = \tfrac{1}{2}\rho \left\{ (\nabla\psi)^2 - \frac{1}{c^2} \left(\frac{d\psi}{dt} \right)^2 \right\}$$

In this simplified system the single field variable ψ is the velocity potential defined by $\mathbf{v} = \nabla\psi$.

With material systems the usefulness of the analytical formulation lies principally in the ease with which transformations may be made to non-cartesian co-ordinate systems suitable for the solution of specific problems. This, of course, confines attention to the Lagrangian formalism. Some use has also been made of the Hamiltonian formalism, mainly in connection with investigating the quantum properties of continuous material systems. A notable case

---

* For a discussion of this aspect see particularly the article by E. L. Hill, *Rev. Mod. Phys.* **23**, 253–60 (1951).

† See the account in the work by Morse and Feshbach quoted in the bibliography.

is the hydrodynamical one where some progress has been made in describing the motion of non-viscous fluids in terms of quantized vibrational motions (phonons) and quantized rotational motions (rotons).

Dissipative processes such as diffusion and friction are of great importance in the study of continuous systems. As with discrete systems, such processes are not easy to incorporate in the analytical description, but the method of the introduction of a complementary 'mirror image' system, described briefly in Chapter V, can be adapted for continuous systems and appears to have interesting possibilities.[*] The Rayleigh dissipation function is also useful if it is not desired to do more than facilitate transformations of the equations of motion to generalized co-ordinates.

The most notable achievements in applying the Lagrangian and Hamiltonian formalisms to continuous systems is in the study of non-material systems known as *fields*. An additional factor then to be taken into account is an emphasis on relativistic invariance. It is found, however, that the theory developed here can be taken over virtually unchanged. This will form the subject of Chapter XI.

* Morse and Feshbach, loc. cit.

CHAPTER X

Relativistic Mechanics

General Framework

The Newtonian system of mechanics is generally accepted as an approximation valid only if the velocities of the particles of the system are small compared to the velocity of light. A more general description is provided by the theory of special relativity. The object now will be to show how the requirements of relativity may be fitted into the Lagrangian and Hamiltonian descriptions. Since the bases of the special theory are adequately dealt with in two other volumes of this series* only a short summary will be given here.

The special theory postulate requires that the laws of nature have the same form in all frames of reference in uniform relative motion. The requirement is met by the Newtonian scheme if attention be confined to purely mechanical systems, but modifications are needed if electromagnetic phenomena are to be included.

A consequence of the relativity principle is that there is no longer a unique time scale common to all frames of reference. A point in ordinary space at a certain instant of time is referred to as an *event* and may be specified by four co-ordinates x_1, x_2, x_3 and t. The same event may be identified in any other frame of reference as (x_1', x_2', x_3', t'). If the second frame S' is moving with respect to the first frame S with uniform velocity V, then the relationship between the two sets of co-ordinates is a linear transformation:

$$\left.\begin{array}{ll} x_1' = x_1 & x_2' = x_2 \\ x_3' = \dfrac{x_3 - Vt}{\sqrt{1 - \beta^2}} & t' = \dfrac{t - Vx_3/c^2}{\sqrt{1 - \beta^2}} \end{array}\right\} \qquad (10.1)$$

where $\beta = V/c$ and c is the velocity of light.

It is assumed that the origins of the two systems coincide initially,

* See the works by Dingle and McCrea quoted in the bibliography.

that their space axes have the same orientations and that the relative motion is along the x_3 direction. Equations (10.1) are known as the *Lorentz transformation relations*.

A more elegant method of description, due to Minkowski, labels an event with the four co-ordinates $(x_1, x_2, x_3, x_4 = ict)$. The four quantities x_μ form the components of a first order four-dimensional cartesian tensor or 4-vector* and the Lorentz transformation relations represent an orthogonal (i.e., length preserving) transformation of such components. It follows that:

$$\sum_\mu x_\mu{}^2 = \sum_\mu x_\mu{}'^2 \tag{10.2}$$

The *interval* or *proper time* between two events x and y is defined as:

$$\tau = \frac{1}{c} \sqrt{\left\{ - \sum_\mu (x_\mu - y_\mu)^2 \right\}} \tag{10.3}$$

τ is an invariant quantity or scalar since it represents the length of a 4-vector. The motion of a particle can be considered as a continuous succession of events forming a curve in four-dimensional Minkowski space. The invariant infinitesimal interval between neighbouring events on this curve is:

$$d\tau = \frac{1}{c} \sqrt{\left\{ - \sum_\mu (dx_\mu)^2 \right\}} = dt \sqrt{\left\{ 1 - \frac{1}{c^2} \sum_i \dot{x}_i{}^2 \right\}} \\ = dt \sqrt{1 - \beta^2} \tag{10.4}$$

where $\beta c = \sqrt{\sum_i \dot{x}_i{}^2} = v$ denotes the ordinary velocity of the particle.

In contrast to the position in Newtonian mechanics the quantities $\dot{x}_i = \dfrac{dx_i}{dt}$ are no longer vector components. This is because t itself is essentially the component of a vector rather than a scalar. (The $\dot{x}_i$ in

* Greek letters will denote that a suffix can range over the values 1, 2, 3, 4. Latin letters will be reserved for the range 1, 2, 3.

fact form three of the sixteen components of a second order tensor.) A genuine 4-vector may be defined by:

$$u_\mu = \frac{dx_\mu}{d\tau} \tag{10.5}$$

Since $\dfrac{dx_\mu}{d\tau} = \dfrac{1}{\sqrt{1 - \beta^2}} \dfrac{dx_\mu}{dt}$ it is obvious that the space-like components of this 4-vector coincide with the ordinary velocity components in the non-relativistic limit when $\beta \rightarrow 0$. This is taken to justify regarding u_μ as the true generalization of the velocity vector. The time-like component is:

$$u_4 = \frac{dx_4}{d\tau} = \frac{ic}{\sqrt{1 - \beta^2}} \tag{10.6}$$

and the invariant length of the velocity 4-vector is:

$$\sum_\mu u_\mu^2 = -c^2 \tag{10.7}$$

It is postulated that each particle has associated with it an invariant quantity m, termed its mass. This being so, it would seem natural to regard the quantities:

$$p_\mu = mu_\mu \tag{10.8}$$

as the components of the 4-vector generalization of the ordinary momentum vector. This is confirmed by arguments based on the assumption that the laws of conservation of mass and of momentum of non-relativistic mechanics are special forms of more general conservation laws. In their more general form these laws state that all four components of the generalized momentum are conserved. The three space-like components correspond to ordinary momentum components, the time-like component:

$$p_4 = \frac{imc}{\sqrt{1 - \beta^2}} \tag{10.9}$$

can be identified with energy, as will be seen later.

Newton's Laws

The use of a tensor notation is primarily to simplify transformations from one frame of reference to another. Within any one frame there seems no point in superseding entirely the more familiar notation used for so long to correlate the facts of observation. In particular there appears to be no reason to dethrone Newton's laws as the basis of the dynamical behaviour of particles. Forces have not, so far, been mentioned in connection with relativity, but it will now be assumed that they can be *defined* by:

$$force = time\ rate\ of\ change\ of\ momentum \qquad (10.10)$$

using the more general definition of momentum given by (10.8),

i.e.,
$$F_\mu = \frac{d}{dt}(p_\mu) = \frac{d}{dt}(mu_\mu) \qquad (10.10')$$

As with the quantities $\dot{x}_\mu = \frac{dx_\mu}{dt}$ this new definition does not give a true 4-vector. There are four quantities F_μ, but the main consideration will be of the three space-like components given explicitly by:

$$F_i = \frac{d}{dt}(p_i) = \frac{d}{dt}\left(\frac{m\dot{x}_i}{\sqrt{1-\beta^2}}\right) \qquad (10.11)$$

It is readily seen that these reduce to the more familiar non-relativistic force components in the low velocity limit. Kinetic energy will also be defined as before through the relation:

$$\frac{dT}{dt} = \mathbf{F}.\mathbf{v} = \sum_i F_i \dot{x}_i \qquad (10.12)$$

A development of this relation yields:

$$\frac{dT}{dt} = \sum_i u_i \sqrt{1-\beta^2}\frac{d}{dt}(mu_i) = \sqrt{1-\beta^2}\left\{\sum_\mu u_\mu\frac{d}{dt}(mu_\mu) - u_4\frac{d}{dt}p_4\right\}$$

$$= \sqrt{1-\beta^2}\frac{d}{dt}\left\{\sum_\mu \tfrac{1}{2}mu_\mu^2\right\} - ic\frac{dp_4}{dt} = -ic\frac{dp_4}{dt} \qquad (10.13)$$

This gives the identification:

$$p_4 = \frac{i}{c}(T + const.) \qquad (10.14)$$

Referring back to (10.9) and postulating that the kinetic energy is zero for the particle at rest, the value of the constant is seen to be mc^2. Hence (10.14) becomes:

$$p_4 = \frac{i}{c}(T + mc^2) = \frac{iE}{c} \tag{10.15}$$

where E, termed the *total energy*, is the sum of the kinetic energy T and the *equivalent mass energy* mc^2.

It now follows that:

$$-m^2c^2 = \sum_\mu p_\mu{}^2 = \sum p_i{}^2 + p_4{}^2 = \mathbf{p}^2 - E^2/c^2$$

i.e., $$(T + mc^2)^2 = E^2 = m^2c^4 + \mathbf{p}^2c^2 = \frac{m^2c^4}{(1 - \beta^2)} \tag{10.16}$$

giving a general relation between the energy and the 3-momentum.

Lagrangian Formulation

If the results so far were to be consistent with a Lagrangian formulation exactly as developed for the non-relativistic case it would be required that:

$$p_i = \frac{\partial L}{\partial \dot{x}_i}$$

i.e., $$\frac{m\dot{x}_i}{\sqrt{1 - \beta^2}} = \frac{\partial L}{\partial \dot{x}_i} \tag{10.17}$$

since $c^2\beta^2 = \sum_i \dot{x}_i{}^2$ this integrates to give:

$$L = -mc^2\sqrt{1 - \beta^2} - V(x_i) \tag{10.18}$$

where, for obvious reasons, the constant of integration has been written as $V(x_i)$. The corresponding Lagrange's equations would then be:

$$-\frac{\partial V}{\partial x_i} = \frac{d}{dt}\left(\frac{m\dot{x}_i}{\sqrt{1 - \beta^2}}\right) \quad \left(= \frac{d}{dt}p_i\right) \tag{10.19}$$

Comparison with (10.11) shows that these are the correct equations of motion if $F_i = -\dfrac{\partial V}{\partial x_i}$. Unfortunately it is difficult to test the correspondence any further since it is not easy to identify any real forces with this prescription. The obvious candidates for identification are gravitational forces, but these are found to require a description outside the considerations of special relativity.

The identification does hold in the special case of a free particle acted upon by no forces. In this case it is possible to describe the behaviour in Lagrangian terms with a Lagrangian function given by (10.18) with $V = 0$. It is also possible to extend the description to a system of free particles but not to a rigid body. The concept of the latter is incompatible with the special relativity postulate since it would demand the existence of interactions between particles which are propagated instantaneously.

Since special relativity theory was formulated explicitly to include electromagnetic phenomena within a general invariant scheme, it is reasonable to hope that the Lagrangian description may also be extended to include these as it did in the non-relativistic case.

The force acting on a particle with charge e is given by the Lorentz expression:

$$\mathbf{F} = e\left(\mathbf{E} + \frac{1}{c}\mathbf{v} \wedge \mathbf{B}\right) \tag{3.16}$$

and it was previously shown that this can be put in the alternative form:

$$F_i = e\left(\frac{d}{dt}\frac{\partial}{\partial \dot{x}_i} - \frac{\partial}{\partial x_i}\right)\left(\phi - \frac{1}{c}\mathbf{v}.\mathbf{A}\right) \tag{3.18}$$

These expressions are based on observations made at low particle velocities and there is no guarantee that they will be generally applicable. If they do hold good at all velocities, then the relativistic equations of motion of a particle moving in an electromagnetic field would be:

$$\frac{d}{dt}\left(\frac{m\dot{x}_i}{\sqrt{1 - \beta^2}}\right) = e\left(\frac{d}{dt}\frac{\partial}{\partial \dot{x}_i} - \frac{\partial}{\partial x_i}\right)\left(\phi - \frac{1}{c}\mathbf{v}.\mathbf{A}\right) \tag{10.20}$$

Further observations using high energy particles show agreement between actual behaviour and predictions based on (10.20), with the identification $v_i = \dot{x}_i$. It is thus concluded that (10.20) are the correct equations of motion and it is not difficult to see, from previous considerations, that they may be derived from the assumption of a Hamilton's principle:

$$\delta \int L \, dt = 0$$

with:
$$L = - mc^2\sqrt{1 - \beta^2} - e\left(\phi - \frac{1}{c}\sum_i \dot{x}_i A_i\right) \qquad (10.21)$$

An extension to a system of such particles presents difficulties owing to the necessity for taking account of the electromagnetic interactions between the particles.

Hamiltonian Formulation

Using the definition $p_i = \dfrac{\partial L}{\partial \dot{x}}$, with L given by (10.21), the space-like components of the momentum of a particle in an electromagnetic field are:

$$p_i = \frac{m\dot{x}_i}{\sqrt{1 - \beta^2}} + \frac{e}{c}A_i \qquad (10.22)$$

Nothing can be said immediately about a time-like component since this is not envisaged in the Lagrangian scheme. The Hamiltonian

$$H = \sum_i p_i \dot{q}_i - L = \sum_i p_i \frac{\sqrt{1 - \beta^2}}{m}\left(p_i - \frac{e}{c}A_i\right)$$
$$+ mc^2\sqrt{1 - \beta^2} + e\left(\phi - \frac{1}{c}\mathbf{v}.\mathbf{A}\right)$$
$$= c\sqrt{\left\{m^2c^2 + \left(\mathbf{p} - \frac{e}{c}\mathbf{A}\right)^2\right\}} + e\phi \qquad (10.23)$$
$$= \frac{mc^2}{\sqrt{1 - \beta^2}} + e\phi \qquad (10.23')$$

As before, the rate of increase of the kinetic energy of the particle is given by:

$$\frac{dT}{dt} = \sum_i \dot{x}_i \frac{d}{dt}p_i = \sum_i \dot{x}_i \frac{d}{dt}\left(\frac{m\dot{x}_i}{\sqrt{1 - \beta^2}}\right) = \frac{d}{dt}\left(\frac{mc^2}{\sqrt{1 - \beta^2}}\right) \qquad (10.24)$$

Again stipulating that $T = 0$ for $\dot{x}_i = 0$, this gives:

$$T = mc^2 \left\{ \frac{1}{\sqrt{1 - \beta^2}} - 1 \right\} \tag{10.24'}$$

From (10.20) and (10.24):

$$\frac{dT}{dt} = \sum_i e\dot{x}_i \left(\frac{d}{dt} \frac{\partial}{\partial \dot{x}_i} - \frac{\partial}{\partial x_i} \right) \left(\phi - \frac{1}{c} \mathbf{v}.\mathbf{A} \right)$$

$$= e\mathbf{v}.\left(\mathbf{E} + \frac{1}{c}\mathbf{v} \wedge \mathbf{B} \right) = e\mathbf{v}.\mathbf{E} \tag{10.25}$$

In terms of the potential functions:

$$\mathbf{E} = -\nabla\phi - \frac{1}{c} \frac{\partial \mathbf{A}}{\partial t}$$

$$\therefore \frac{dT}{dt} = -e\mathbf{v}.\left(\nabla\phi + \frac{1}{c} \frac{\partial \mathbf{A}}{\partial t} \right) = -\frac{d}{dt}(e\phi) + e\frac{\partial \phi}{\partial t} - \frac{e}{c}\mathbf{v}\cdot\frac{\partial \mathbf{A}}{\partial t}$$

i.e.,

$$\frac{d}{dt}(T + e\phi) = e\left(\frac{\partial \phi}{\partial t} - \frac{1}{c}\mathbf{v}\cdot\frac{\partial \mathbf{A}}{\partial t} \right)$$

or:

$$\frac{d}{dt}(mc^2 + T + e\phi) = e\left(\frac{\partial \phi}{\partial t} - \frac{1}{c}\mathbf{v}\cdot\frac{\partial \mathbf{A}}{\partial t} \right) \tag{10.26}$$

The quantity $(mc^2 + T + e\phi) = E$ may be interpreted as the total energy of the particle. The meaning of the right-hand side terms of (10.26) is, then, that they represent an increase in the total energy of the particle due to the action of the charges and currents which are causing the field strength to vary with time. With this interpretation the Hamiltonian is identical with the total energy, as will be seen by combining (10.23') and (10.24'). Note, however, that the Lagrangian function is not to be equated with any expression such as $(T + mc^2 - e\phi)$, as might be inferred from the non-relativistic case.

Equation (10.23), giving the Hamiltonian in its canonical form, serves as a starting-off point for Dirac's relativistic quantum theory of the electron.

It has now been demonstrated that both Lagrangian and Hamiltonian formulations can be found which describe the behaviour of single charged particles in an electromagnetic field. As previously mentioned, an extension to a system of particles, or to behaviour under other influences such as that of a gravitational field, is by no means simple.

A Covariant Formulation

An alternative formulation to that of the previous section can be found which places more emphasis on tensor notation. The starting-point is the realization that the scalar quantity τ (the proper time) could be regarded as a generalized form of the variable t since the latter behaves as a scalar in non-relativistic mechanics. A generalized form of Hamilton's principle would then be:

$$\delta \int L \, d\tau = 0 \tag{10.27}$$

where, if L is a scalar, the principle is identically the same for all frames of reference. Such a formulation is said to be *covariant* in form.

It is not difficult to see that the choice:

$$L = \sum_\mu \tfrac{1}{2} m u_\mu{}^2 \tag{10.28}$$

is a function of the required form and that it meets the requirements made of a Lagrangian used to describe the motion of a free particle of mass m.

It must now be accepted that there are four variables x_μ which are to have equal weight in the formulation and there will therefore be four equations of motion of the form:

$$\frac{d}{d\tau}\left(\frac{\partial L}{\partial x_\mu{}'}\right) - \frac{\partial L}{\partial x_\mu} = 0 \tag{10.29}$$

where $x_\mu{}' \equiv \dfrac{dx_\mu}{d\tau}$. From (10.28) these equations would be:

$$\frac{d}{d\tau}(m u_\mu) = 0 \tag{10.30}$$

the three equations corresponding to the ordinary space variables reduce to:

$$\frac{d}{dt}\left(\frac{m\dot{x}_i}{\sqrt{1 - \beta^2}}\right) = 0 \tag{10.31}$$

which have already been accepted as correct. The fourth equation becomes:

$$\frac{d}{dt}\left(\frac{imc}{\sqrt{1-\beta^2}}\right) = 0 \qquad (10.32)$$

which expresses the constancy of the total energy.

The momentum components will be also now four in number:

$$p_\mu = \frac{\partial L}{\partial x_\mu{}'} = mu_\mu \qquad (10.33)$$

These are again identical with results previously agreed.

For consistency the Hamiltonian must now be defined as:

$$H = \sum_\mu p_\mu x_\mu{}' - L \qquad (10.34)$$

where the summation is over four, not three, terms. This gives:

$$H = \sum_\mu \frac{p_\mu^2}{2m} = -\tfrac{1}{2}mc^2 \qquad (10.35)$$

which is *not* identical with the total energy $\left(E = \dfrac{mc^2}{\sqrt{1-\beta^2}}\right)$. The canonical equations:

$$\frac{d}{d\tau}(p_\mu) = -\frac{\partial H}{\partial x_\mu} \quad \text{and} \quad \frac{d}{d\tau}(x_\mu) = \frac{\partial H}{\partial p_\mu}$$

do, however, give valid relations.

It might be thought that the non-identity of the Hamiltonian and total energy would be a serious objection to this formulation. This is not the case since the latter may be obtained from the 4-component of the momentum using the identification made earlier in this chapter:

$$p_4 = \frac{iE}{c} \qquad (10.15)$$

The covariant formulation, unlike that of the previous sections, determines the energy as a derivative of the Lagrangian.

In turning to a consideration of the motion of a particle in an electromagnetic field mention must now be made of a restriction which has to be placed on the potential functions A and ϕ. There is a certain degree of freedom in the definitions of these quantities which can be removed in various ways. In relativistic considerations it is convenient to choose:

$$\nabla.\mathbf{A} + \frac{1}{c}\frac{\partial \phi}{\partial t} = 0 \qquad (10.36)$$

It was unnecessary to introduce this into earlier considerations but it now enables the identification of a new 4-vector. (10.36) may be rewritten as:

$$\frac{\partial A_i}{\partial x_i} + \frac{\partial (i\phi)}{\partial x_4} = 0 \qquad (10.36')$$

from which it can be inferred that the four quantities $(A_1, A_2, A_3, A_4 = i\phi)$ are the components of a 4-vector. (10.36') merely states that the (4)-divergence of this vector is zero.

Bearing this new vector quantity in mind and referring back to the non-covariant Lagrangian function for the particle in the field, it would seem plausible that a suitable covariant Lagrangian function might take the form:

$$L = \sum_{\mu} \left\{ \tfrac{1}{2}mu_{\mu}^2 + \frac{e}{c}A_{\mu}u_{\mu} \right\} \qquad (10.37)$$

This is a scalar quantity; thus the first requirement is fulfilled. The corresponding equations of motion would be:

$$\frac{d}{d\tau}\left(mu_{\mu} + \frac{e}{c}A_{\mu} \right) = \sum_{\nu} \frac{e}{c}u_{\nu}\frac{\partial A_{\nu}}{\partial x_{\mu}} \qquad (10.38)$$

these may be reduced to:

$$\frac{d}{dt}\left(\frac{m\mathbf{v}}{\sqrt{1-\beta^2}} \right) = e\left(\mathbf{E} + \frac{1}{c}\mathbf{v} \wedge \mathbf{B} \right) \qquad (10.39)$$

and

$$\frac{d}{dt}\left(\frac{mc^2}{\sqrt{1-\beta^2}} \right) = e\mathbf{v}.\mathbf{E} \qquad (10.40)$$

Equations (10.39) have been previously quoted as correct. (10.40), corresponding to the time co-ordinate x_4, is identical with (10.25). The proposed Lagrangian function thus provides the correct set of equations of motion.

The momentum components deducible from (10.37) are:

$$p_\mu = mu_\mu + \frac{e}{c}A_\mu \tag{10.41}$$

The first three of these equations agree with (10.22) and are therefore acceptable, the fourth has not previously appeared:

$$p_4 = mu_4 + \frac{e}{c}A_4 = \frac{i}{c}\left(\frac{mc^2}{\sqrt{1-\beta^2}} + e\phi\right) \tag{10.42}$$

The bracketed quantity was, in the discussion following (10.26), identified as the total energy of the particle; hence the same equality holds as in the free particle case:

$$p_4 = iE/c \tag{10.43}$$

The Hamiltonian is:

$$\left.\begin{aligned} H &\equiv \sum_\mu p_\mu x_\mu' - L = \sum_\mu \frac{1}{2m}\left(p_\mu - \frac{e}{c}A_\mu\right)^2 \\ &= \sum_\mu \tfrac{1}{2}mu_\mu^2 = -\tfrac{1}{2}mc^2 \end{aligned}\right\} \tag{10.44}$$

This again is not identifiable with the total energy. But, as before, this is of little consequence in view of (10.43). The covariant formulation thus forms a rather elegant method for the determination of the physically interesting relations (equations of motion) and quantities (momentum and energy). In adopting it, an almost complete break is made with the Newtonian scheme since the notion of force has now completely disappeared. It is, of course, possible to identify force components by translating the equations of motion back into the ordinary space-time notation, but there are no simple tensor quantities to which they correspond.

The results of this section provide an essentially parallel formulation, in Lagrangian and Hamiltonian terms, to that of the previous sections. It should be noted, however, that the covariant formulation involves much simpler fundamental expressions. Given that it has to depend upon such quantities as x_μ, u_μ and A_μ, the Lagrangian function (10.37) is one of the simplest scalar combinations that can be devised. This aspect has proved a profitable hint in the study of more complicated systems. The next chapter will deal with the elementary aspects of field theories in this connection.

CHAPTER XI

Fields

A prominent feature of modern physics is the large number of different types of *fundamental particle* which have been observed and identified. In addition to the familiar *electrons, protons, neutrons* and *photons* there is now a formidable list of other types of particles.* Much investigation remains to be carried out before the properties of even the more well known types of particle are satisfactorily classified; nevertheless, considerable success has already been achieved in developing a theoretical description.

The most outstanding characteristic of all types of particle so far discovered is that in some ways they behave as particles in the classical sense and in others they appear to be associated with some form of wave motion. Historically it was the particle aspects of ordinary matter that were first studied; the wave properties, described by quantum theory, were considered much later. In the case of photons the reverse development occurred; the theory of the electromagnetic field was developed, then came the realization that certain properties of electromagnetic waves could only be explained by postulating the existence of discrete, particle-like, entities termed photons.

This wave-particle duality is now accepted as a general feature of the behaviour of all types of fundamental particle. In the development of the theoretical description it has been found generally most productive to develop first the wave description and then to proceed to the particle aspects. The wave description requires the development of a *field theory*. This is in classical terms. At a certain stage quantum rules are incorporated in the description, and it is then found possible to interpret some of the deductions in terms of particle concepts.

* See, for instance, *Introduction to Elementary Particle Physics* by R. E. Marshak and E. C. G. Sudarshan, Interscience, New York, 1961.

The study of field theories has been developed as a logical extension to that of continuous material systems. In this chapter it is proposed to give a brief account of the essential features of such theories. The quantum aspects will, of course, lie outside our considerations, though the procedure for introducing the quantum rules is basically as indicated in Chapter IX.

The best known example of the type of system under consideration is the electromagnetic field. This may be described in terms either of the electric and magnetic field strengths or of the scalar and vector potential functions; in either case the quantities concerned are continuously variable functions of space and time. This form of description is based ultimately upon observations of the motions of ordinary material particles postulated to carry electric charges. The idea of a continuous *field* is introduced in order to avoid the concept of 'action at a distance' between the particles. The sources of the field are the charges residing on the particles. The idea is refined and idealized to the extent that the field is considered to exist in some form even in the absence of the particles. The properties of the electromagnetic field are summarized in the set of differential relations known as Maxwell's equations. These will usually be referred to as the *field equations*.

It is assumed that fields are associated with other types of fundamental particle in the same way that the electromagnetic field is associated with photons. These do not necessarily possess the same degree of complexity as the electromagnetic field; indeed, some are much simpler. The fundamental assumption made is that the wave-like behaviour of any type of particle may be summarized in a set of field equations involving one or more field variables. It is also assumed that the equations must be invariant to Lorentz transformations* and thus conform to the relativistic requirement that all the basic laws of nature shall take the same form in all frames of reference. This is already the case with Maxwell's equations, although the development of electromagnetic theory preceded that of special relativity.

Field variables are not accessible to direct observation but their

* In this simplified account only proper Lorentz transformations will be considered.

values, as in the electromagnetic case, may be deduced from observations on material systems. The realization of this fact should obviate queries as to the *nature* of these variables. They have as much, or as little, reality as electric or magnetic field vectors, or even the potential functions of classical mechanics. They are best regarded as mathematical entities whose significance lies in the possibility of their being used to describe and predict observable changes in the behaviour, ultimately, of material systems.

The field theory describing the relativistic wave properties of ordinary matter involves rather complicated entities known as *spinors*. Any consideration of spinor fields would carry us beyond the scope of this book and it will be necessary to use much simpler systems for the purposes of illustration. This may possibly lead to an air of unreality, but it is thought preferable to consider simple and, even at times, hypothetical, examples which readily convey the principles involved, rather than to attempt a considerably more difficult analysis in which these principles may get submerged. In this way it is hoped to provide a simple introduction to the general lines of development of field theories. For detailed treatment of the subject the reader is referred to the treatises listed in the bibliography.

Lagrangian Formulation

The formulation developed in Chapter IX, although designed specifically for continuous material systems, forms such a compact description as to suggest that it might usefully be taken over for the description of fields. No restriction was there placed on the nature of the field variables and they could just as well be the potential functions of the electromagnetic field as the displacements of an elastic solid or the velocity potential of a fluid. There is little doubt that suitable density functions can be found which summarize the properties of fields, the point to be investigated is whether they lead to any essential simplification.

The general situation in field theory is somewhat different from that with regard to continuous material systems. In the latter type of system the fundamental behaviour is usually fairly well understood and the analytical formulation is used to simplify the process of writing down the equations of motion in a form suited to the solution of specific problems. With fields, previous knowledge of the basic behaviour is usually non-existent, and the analytical method is used as the starting-point of the theoretical description. The consideration of various types of simple Lagrangian density function has led to an encouraging degree of success in explaining some of the observed phenomena. The analytical approach is no less empirical than the one in which direct guesses are made concerning the form of the field equations, but the range of possibilities is considerably reduced.

It was found that the properties of continuous systems can be summarized in an extended form of Hamilton's principle:

$$\delta \int \int \mathscr{L} \, dV \, dt = 0 \tag{9.12}$$

where, in general, $\mathscr{L} = \mathscr{L}(\eta^{(r)}, \dot{\eta}^{(r)}, \eta, {}_i^{(r)}, x_i, t)$.

The requirement that field theories should be sufficiently general to cover the requirements of special relativity has already been mentioned. In connection with the study of particle motion in analytical terms in Chapter X it was found possible to incorporate relativity effects in two ways. Of these, the covariant method was by far the simpler. This is taken as a guide to present procedure in the case of fields. It is not possible to adopt the covariant formulation exactly as in Chapter X, but a re-examination of (9.12) suggests an alternative. Since $dV = dx_1 \, dx_2 \, dx_3 \, (\equiv d^3x)$ and $x_4 = ict$, this equation can be rewritten:

$$\delta \int \mathscr{L} d^4x \equiv \delta \int \int \int \int \mathscr{L} \, dx_1 \, dx_2 \, dx_3 \, dx_4 = 0 \tag{11.1}$$

where the constant factor ic has been absorbed in $\mathscr{L}$.

The infinitesimal product $d^4x \equiv dx_1 \, dx_2 \, dx_3 \, dx_4$ represents an element of volume in Minkowski 4-space and, as such, is invariant to Lorentz transformations. It follows that (11.1) is itself invariant provided only that $\mathscr{L}$ is a scalar quantity. This is the form of Hamilton's principle that is taken over as a starting-point for the description of fields. Since it is merely a rewritten version of the principle used previously, all the former consequences also apply here. In particular, the equations of motion (translated into the new notation) will be:

$$\sum_{\mu} \frac{d}{dx_{\mu}} \frac{\partial \mathcal{L}}{\partial \eta_{,\,\mu}^{(r)}} - \frac{\partial \mathcal{L}}{\partial \eta^{(r)}} = 0 \qquad (11.2)$$

These equations are now to be referred to as the field equations. Their invariant form with respect to Lorentz transformations is guaranteed by the invariance of Hamilton's principle in the form (11.1).

Types of Field

(a) *Scalar Fields.* The simplest conceivable type of field would be that having a single real scalar field variable ϕ. Confining attention to derivatives of ϕ no higher than the first, obvious scalar quantities which can be constructed from it comprise integral powers of ϕ and of $\sum_{\mu} \left(\dfrac{d\phi}{dx_{\mu}} \right)^2$.

A simple Lagrangian density function based on such quantities would be:*

$$\mathcal{L} = \alpha \left\{ \sum_{\mu} \left(\frac{d\phi}{dx_{\mu}} \right)^2 + k^2 \phi^2 \right\} \qquad (11.3)$$

From (11.2) the associated equations of motion would then take the form:

$$\sum_{\mu} \frac{d^2 \phi}{dx_{\mu}^2} - k^2 \phi = 0$$

i.e., $(\Box - k^2)\phi = 0 \qquad (11.4)$

where $\Box \equiv \nabla^2 - \dfrac{1}{c^2} \dfrac{d^2}{dt^2}$ is the d'Alembertian operator.

From arguments involving quantum theory considerations, such a field can be shown to describe the wave aspects of neutral particles having a rest mass $m = kh/2\pi c$. Neutral π mesons are thought to be associated with a related type of pseudoscalar field differing only in that ϕ changes sign under space inversion, i.e. $\phi(\mathbf{r},\, t) = -\phi(-\mathbf{r},\, t)$. The effect of this modification is apparent only in interactions with other fields.

* See the footnote on page 98 regarding the convention adopted in the use of differentiation symbols.

Other possible types of scalar field have also been investigated. It can, for instance, be shown that a complex scalar field variable $\phi = \phi_1 + i\phi_2$ is associated with electrically charged particles.

(b) *Vector Fields.* A more complicated type of field would have as field variables the components of a 4-vector. The electromagnetic field is, in fact, an example of this type of field.

The electromagnetic field in the absence of matter is described by Maxwell's equations for free space, with charge and current densities everywhere zero:

i.e.,
$$\nabla . \mathbf{B} = 0 \qquad \nabla \wedge \mathbf{E} = -\frac{1}{c}\frac{d\mathbf{B}}{dt} \qquad (11.5a)$$

$$\nabla . \mathbf{E} = 0 \qquad \nabla \wedge \mathbf{B} = \frac{1}{c}\frac{d\mathbf{E}}{dt} \qquad (11.5b)$$

The defining relations for the vector and scalar potential functions are:

$$\mathbf{B} = \nabla \wedge \mathbf{A} \qquad \mathbf{E} = -\nabla\phi - \frac{1}{c}\frac{d\mathbf{A}}{dt} \qquad (11.6)$$

In terms of these definitions (11.5a) are mere identities and will not be considered further. This leaves equations (11.5b) as the field equations proper.

As pointed out in Chapter X, the four quantities (A_1, A_2, A_3, A_4 = $i\phi$) may be identified as the components of a 4-vector from the relation:

$$\nabla . \mathbf{A} + \frac{1}{c}\frac{d\phi}{dt} = 0$$

this reduces the arbitrariness in the definitions (11.6)* and confines attention to the so-called Lorentz gauge. Using this gauge, the equations (11.5b) may be rewritten in the relativistically invariant form:

$$\sum_{\mu}\frac{d}{dx_\mu}\left(\frac{dA_\mu}{dx_\nu} - \frac{dA_\nu}{dx_\mu}\right) = 0 \qquad (11.7)$$

---

* The A_μ remain arbitrary to the extent of additive quantities $\dfrac{d\Lambda}{dx_\mu}$ where $\square\Lambda = 0$. This is of no consequence. It is $\mathbf{E}$ and $\mathbf{B}$ that are required to be fully defined.

alternatively, if we define an antisymmetric second order tensor $F_{\mu\nu}$ by:

$$F_{\mu\nu} = \frac{dA_\mu}{dx_\nu} - \frac{dA_\nu}{dx_\mu} \tag{11.8}$$

the field equations take the form:

$$\sum_\mu \frac{dF_{\mu\nu}}{dx_\mu} = 0 \tag{11.7'}$$

It is not difficult to see that these equations may be deduced from an assumption of a Hamilton's principle of the form (11.1) with:

$$\mathscr{L} = \alpha \sum_\mu \sum_\nu (F_{\mu\nu})^2 = \alpha \sum_{\mu,\ \nu} \left(\frac{dA_\mu}{dx_\nu} - \frac{dA_\nu}{dx_\mu} \right)^2 \tag{11.9}$$

A word of warning should here be given that, although this simple form of Lagrangian gives the correct field equations, it is unsatisfactory on other grounds. Modifications are, however, only necessary when going outside the Lagrangian formalism and will not be considered here.

Other types of vector field have also been considered. In particular it has been shown that, as in the scalar case, complex field components imply that the associated particles are electrically charged. A further generalization, based on non-classical considerations, is that the particles associated with any vector field have an intrinsic angular momentum of amount $h/2\pi$ (unit spin). This is in contrast to scalar fields which are associated with particles of zero spin.

(*c*) *Spinor Fields.* The wave properties of ordinary matter have also been described in terms of a field theory. The description involves specialized entities, known as spinors, as field variables. It is not intended to consider these further but it may be remarked that the associated particles possess half-integral spin values. This is, of course, in agreement with the facts of observation.

In any field theory the type of field variable always determines the spin characteristic of the associated particles. In any particular case this again narrows the range of possible Lagrangian density functions. Some consideration has been paid to fields with particles of spin other than 0, $\frac{1}{2}$ and 1, but the most important ones are those considered above. In constructing Lagrangian densities it is usual to confine consideration to functions which contain the various field quantities to second or lower powers. This accords with the usual experience that field equations are at most second order differential equations.

Field Interactions

So far only individual 'free' fields have been considered. This meets only part of the final requirement since fields interact with each other. This fact is demonstrated by the appearance of terms involving current and charge densities in the more general form of Maxwell's equations, indicating ultimately an interaction between the electromagnetic field and the spinor field associated with ordinary matter. Without such interactions with the matter field it would, in fact, be impossible to observe any properties of other types of field.

To incorporate interactions into the description it is assumed that two interacting fields A and B can be described by a Lagrangian density function of the form:

$$\mathcal{L} = \mathcal{L}^{(A)} + \mathcal{L}^{(B)} + \mathcal{L}^{(int)}$$

where $\mathcal{L}^{(A)}$ and $\mathcal{L}^{(B)}$ relate to the isolated 'free' fields and $\mathcal{L}^{(int)}$ describes their interaction properties. $\mathcal{L}^{(int)}$ is, in general, a scalar function of both sets of field variables. The general case may be illustrated by the following purely hypothetical example:

Suppose that A is a scalar field with one variable ϕ whose properties are deducible from the Lagrangian density function $\mathcal{L}^{(A)} = \sum_{\mu} \left(\dfrac{d\phi}{dx_{\mu}}\right)^2 + k^2\phi^2$ and B is a vector field with a Lagrangian $\mathcal{L}^{(B)} = \sum_{\nu, \mu} \left(\dfrac{d\psi_{\nu}}{dx_{\mu}}\right)^2$. Then the properties of the combined system will be summarized in:

$$\mathcal{L} = \sum_{\mu} \left(\frac{d\phi}{dx_{\mu}}\right)^2 + k^2\phi^2 + \sum_{\nu, \mu} \left(\frac{d\psi_{\nu}}{dx_{\mu}}\right)^2 + \mathcal{L}^{(int)} \quad (11.10)$$

$\mathscr{L}^{(int)}$ will be a scalar function of ϕ and the ψ_ν. Two simple possibilities are:

$$\mathscr{L}^{(int)} = g \sum_\mu \psi_\mu \frac{d\phi}{dx_\mu} \qquad (11.11a)$$

and

$$\mathscr{L}^{(int)} = g'\phi \sum_\mu (\psi_\mu)^2 \qquad (11.11b)$$

where the g's are independent of the field variables and are termed *coupling constants*. (11.11a) would be said to represent derivative coupling and (11.11b) direct coupling between the two fields.

Assuming derivative coupling represented by (11.11a), the full Lagrangian becomes:

$$\mathscr{L} = \sum_\mu \left(\frac{d\phi}{dx_\mu}\right)^2 + k^2\phi^2 + \sum_{\nu, \mu} \left(\frac{d\psi_\nu}{dx_\mu}\right)^2 + g \sum_\mu \psi_\mu \frac{d\phi}{dx_\mu} \quad (11.10')$$

Applying the standard procedure the corresponding equations of motion would be:

$$2 \sum_\mu \frac{d^2\phi}{dx_\mu^2} - 2k^2\phi = -g \sum_\mu \frac{d\psi_\mu}{dx_\mu} \qquad \sum_\mu \frac{d^2\psi_\nu}{dx_\mu^2} = g\frac{d\phi}{dx_\nu} \quad (11.12)$$

The effect of the interaction term is thus seen to be to add coupling terms, represented by the right-hand sides of equations (11.12), to what were previously independent wave equations.

Again it must be emphasized that the example quoted is a purely hypothetical one, but it serves to illustrate the general effect. It is generally true that exact solutions can only be found for the equations relating to the pure fields. The more complicated equations for the interacting systems are usually dealt with by some perturbation procedure in which the coupling terms are assumed small. This is an acceptable method of dealing with the interaction between the electromagnetic field and ordinary matter, but in some known cases the coupling constant is so large as to invalidate the procedure. This constitutes a major problem of modern theoretical physics.

Theoretical investigations of the interactions between fundamental particles have largely taken the form of trying out the more obvious types of interaction Lagrangians and this approach has, on the whole, proved fruitful. Among the more complicated possibilities is a simultaneous interaction between more than two fields. It was implied in the above considerations that all the variables in the interaction Lagrangian density refer to the same point of the field, in other words the interaction is a *local* one. Considerations may also be extended to include a more general type of *non-local* interaction. All such extensions must be seriously considered since, in spite of the theoretical successes which have been achieved, the volume of established explanation is still comparatively small. Simple theories are attractive but there is no logical reason to suppose that all phenomena conform to a simple Lagrangian description.

Interaction between the Electromagnetic Field and Matter

As previously emphasized, a full account of the spinor fields associated with ordinary matter is beyond our present scope. It is correspondingly not feasible to consider the interaction between the electromagnetic field and a spinor field. Nevertheless, some light can be shed by considering the interaction in different terms.

A form of Hamilton's principle giving the correct equations for the 'free' electromagnetic field was found to be:

$$\delta \iiiint \sum_{\mu, \nu} \alpha \left(\frac{dA_\mu}{dx_\nu} - \frac{dA_\nu}{dx_\mu} \right)^2 dx_1 \, dx_2 \, dx_3 \, dx_4 = 0 \quad (11.13)$$

Also, a form of Hamilton's principle giving the correct relativistic equations of motion of a charged particle in an electromagnetic field was found in Chapter X to be:

$$\delta \int \left\{ - mc^2 \sqrt{1 - \beta^2} + \sum_\mu \frac{e}{c} \dot{x}_\mu A_\mu \right\} dt = 0 \quad (11.14)$$

It is plausible to combine these two principles to form a single principle describing the behaviour of the combined system:

$$\delta \int \left\{ \frac{1}{ic} \left(- mc^2 \sqrt{1 - \beta^2} + \sum_\mu \frac{e}{c} \dot{x}_\mu A_\mu \right) \right. \\ \left. + \int \alpha \sum_{\mu, \nu} \left(\frac{dA_\mu}{dx_\nu} - \frac{dA_\nu}{dx_\mu} \right)^2 d^3 x \right\} dx_4 = 0 \quad (11.15)$$

This new principle certainly supplies the correct equations for the motion of the particle from the variation of the particle co-ordinates. The result of varying the A_μ can only be determined after first modifying (11.15) in such a way that all the terms involving the A_μ are included under a common integral sign. To this end we put:

$$\frac{e}{c}\dot{x}_\mu A_\mu = \int \frac{e}{c}\dot{x}_\mu A_\mu \delta(\mathbf{r} - \mathbf{r}_0) \, d^3x \tag{11.16}$$

where $\delta(\mathbf{r} - \mathbf{r}_0) \equiv \delta(x_1 - x_1^{(0)})\delta(x_2 - x_2^{(0)})\delta(x_3 - x_3^{(0)})$,
$\mu_0 = (x_1^{(0)}, x_2^{(0)}, x_3^{(0)})$ is the position vector of the particle and $\delta(x - x_0)$ is the Dirac delta function defined by:

$$\delta(x - x_0) = 0 \quad x \neq x_0 \quad \text{and} \quad \int f(x)\delta(x - x_0)dx = f(x_0) \tag{11.17}$$

assuming the region of integration includes x_0.

We further identify $e\delta(\mathbf{r} - \mathbf{r}_0)$ as the charge density ρ (in strict accordance with the foregoing this is infinite at $\mathbf{r} = \mathbf{r}_0$ and zero elsewhere) hence:

$$\int \rho \, d^3x = \int e\delta(\mathbf{r} - \mathbf{r}_0)d^3x = e \tag{11.18}$$

We next put $\rho\dot{x}_\mu = j_\mu$, where j_i represents the three components of the normal current vector and $j_4 = ic\rho$. With these modifications (11.15) becomes:

$$\delta\int\left\{- mc^2\sqrt{1 - \beta^2} \\ + \int\left(\frac{1}{c}\sum_\mu j_\mu A_\mu + ic\alpha\sum_{\mu,\,\nu}\left(\frac{dA_\mu}{dx_\nu} - \frac{dA_\nu}{dx_\mu}\right)^2\right)d^3x\right\}dx_4 = 0 \tag{11.15'}$$

The A_μ may now be varied giving:

$$4ic^2\alpha\sum_\nu \frac{d}{dx_\nu}\left(\frac{dA_\mu}{dx_\nu} - \frac{dA_\nu}{dx_\mu}\right) = j_\mu \tag{11.19}$$

Retranslated into more familiar notation these equations are:

$$\left.\begin{array}{c} \nabla \wedge \mathbf{B} - \dfrac{1}{c}\dfrac{d\mathbf{E}}{dt} = -\dfrac{1}{4\alpha ic^2}\,\mathbf{j} \\ \nabla.\mathbf{E} = -\rho/4\alpha ic \end{array}\right\} \tag{11.20}$$

If $\alpha = -\dfrac{1}{16\pi ic}$ they may be identified with the more general form of Maxwell's equations, valid when there are non-zero current and charge densities.

Since all the relevant equations of motion may be deduced from it, the hybrid Hamilton's principle, given by (11.15), is concluded to be capable of representing the interacting systems. As a result of correlating the deduced with the accepted form of the equations a value has been found for the constant α which was otherwise arbitrary. This is because the whole expression has been made homogeneous, with all terms of the integrand in (11.15′) representing energies or energy densities. This may be seen from the fact that:

$$ic\alpha \sum_{\mu,\,\nu} \left(\frac{dA_\mu}{dx_\nu} - \frac{dA_\nu}{dx_\mu} \right)^2 = -\frac{1}{8\pi}(\mathbf{B}^2 - \mathbf{E}^2) \tag{11.21}$$

in which will be recognized the familiar expressions for the electric and magnetic field energy densities.

The term $\displaystyle\sum_\mu \frac{1}{c} j_\mu A$ may be identified as the interaction Lagrangian density for the case under consideration. It is of the same form as the corresponding term in the fuller treatment in which the interaction is considered as one between the electromagnetic field and a spinor field.

Hamiltonian Formulation

As in Chapter X the starting-point of the Hamiltonian scheme is the definition of conjugate momenta through:

$$\pi^{(r)} = \frac{\partial \mathscr{L}}{\partial \dot{\eta}^{(r)}} = \frac{1}{ic} \frac{\partial \mathscr{L}}{\partial \eta,_4^{(r)}} \tag{11.22}$$

Although the label is still retained, these quantities have now lost all meaning as momenta in the original sense. It is a source of some difficulty that quantities so defined are non-relativistic in that they single out the time co-ordinate in a special way. A solution to this difficulty will be indicated in a later section. For many purposes it is ignored and for the moment we shall content ourselves with the definition (11.22) and translate the Hamilton formalism of Chapter IX into 4-tensor notation.

The Hamiltonian density function is defined as:

$$\mathcal{H} = \sum_r \dot{\eta}^{(r)} \pi^{(r)} - \mathcal{L} = \sum_r \eta,_4^{(r)} \frac{\partial \mathcal{L}}{\partial \eta,_4^{(r)}} - \mathcal{L} \qquad (11.23)$$

The canonical equations of motion follow as before:

i.e., $$\qquad\qquad \dot{\pi}^{(r)} = -\frac{\delta H}{\delta \eta^{(r)}} \qquad \dot{\eta}^{(r)} = \frac{\delta H}{\delta \pi^{(r)}} \qquad\qquad (9.23')$$

It is not, however, worthwhile to translate these into the new notation in view of their essentially non-invariant nature.

The Lagrangian density function summarizing the properties of any given field is always arbitrary to the extent that the 4-divergence of any 4-vector function of the various field quantities may be added to it. This is because the field equations and integral functions such as H which represent the physical content of the theory, are unchanged by such additions.* Such modifications may, on the other hand, alter the functional dependence of the Hamiltonian and other density functions. The resulting degree of freedom is utilized to ensure the symmetry of the stress-energy-momentum tensor as discussed in the next section.

Energy was defined originally in connection with the study of ordinary particle systems; as a consequence the term has no meaning unless it may be referred back to such systems. It follows that no logical reference can be made to the energy of a field until account has been taken of an interaction between the field in question and that associated with ordinary matter. It was essentially this correlation that was carried out in the previous section, when a value was found for the otherwise arbitrary constant appearing in the Lagrangian density function for the electromagnetic field.

Conservation Laws

In company with the rest of the formalism the density conservation

---

* The variation of the extra term appearing under the integral sign in Hamilton's principle is zero. Also such terms integrate to zero as in the discussion of conservation laws in Chapter IX.

laws of Chapter IX may also be shown, with the same restriction, to apply to fields:

i.e.,
$$\frac{d\mathscr{H}}{dt} + \sum_j \frac{dS_j}{dx_j} = 0 \qquad (9.25')$$

$$\frac{d\mathscr{G}_j}{dt} - \sum_i \frac{dT_{ij}}{dx_i} = 0 \qquad (9.29)$$

The quantities concerned may, however, now be combined under one unifying definition:

$$T_{\mu\nu} = \left\{ \sum_r \frac{\partial \mathscr{L}}{\partial \eta_{,\,\mu}{}^{(r)}} \eta_{,\,\nu}{}^{(r)} - \delta_{\mu\nu} \mathscr{L} \right\} \qquad (11.24)$$

since:
$$\left.
\begin{aligned}
T_{4j} &= ic \sum_r \frac{\partial \mathscr{L}}{\partial \dot{\eta}^{(r)}} \eta_{,\,j}{}^{(r)} = -ic\mathscr{G}_j \\[2mm]
T_{i4} &= \frac{1}{ic} \sum_r \frac{\partial \mathscr{L}}{\partial \eta_{,\,i}{}^{(r)}} \dot{\eta}^{(r)} = \frac{1}{ic} S_i \\[2mm]
T_{44} &= \sum_r \frac{\partial \mathscr{L}}{\partial \dot{\eta}^{(r)}} \dot{\eta}^{(r)} - \mathscr{L} = \mathscr{H}
\end{aligned}
\right\} \qquad (11.24')$$

and the (ij) components are identical with those labelled similarly in Chapter IX.

$T_{\mu\nu}$ can be identified as a second order 4-tensor quantity and is usually referred to as the *stress-energy-momentum tensor*. It is readily seen that the four conservation laws may now be summarized in the form of a general 4-divergence relation:

$$\sum_\mu \frac{dT_{\mu\nu}}{dx_\mu} = 0 \qquad (11.25)$$

The 4-divergence of any second order tensor is a vector quantity. (11.25) asserts that in this particular case the vector is zero.

Since $T_{\mu\nu}$ is a tensor quantity it also follows that the four integral quantities $\int T_{4\mu}\,d^3x \equiv \iiint T_{4\mu}\,dx_1\,dx_2\,dx_3$ form the components of a 4-vector. From (11.24') we have:

$$\left.\begin{aligned}
\int T_{4j}\,d^3x &= -ic\int \mathscr{G}_j\,d^3x = -icG_j \\
\int T_{44}\,d^3x &= \int \mathscr{H}\,d^3x = H
\end{aligned}\right\} \tag{11.26}$$

hence the following quantities are identified as the components of a vector:

$$P_\mu = \left(G_1, G_2, G_3, \frac{iH}{c}\right) \tag{11.27}$$

This is analogous to the result for a relativistic particle and the vector concerned is termed the generalized 4-momentum of the field. It is possible to identify it with the total generalized 4-momentum of the associated particles of the field.

It should be noted that the conditions of Chapter IX still apply for the constancy of the quantities P_μ with time; i.e., the independent variables x_μ do not appear explicitly in $\mathscr{L}$ and either the system is finite in extent, with its physical boundaries within the region of integration, or else there is some form of periodic boundary condition.

The conservation of the generalized wave angular momentum may be considered in addition to that of the linear momentum. It may be shown that this requires the tensor $T_{\mu\nu}$ to be symmetric. This is a generalization of the result obtained in Chapter IX and is usually imposed as a condition on the field. In many cases the obvious Lagrangian density function gives rise to a tensor which is already symmetrical. In other cases it may not do so, but it is always possible to remedy the matter by a re-definition involving the addition to the original tensor $T_{\mu\nu}$ of another, $T'_{\mu\nu}$, satisfying the relations:

$$\left.\begin{aligned}
T_{\mu\nu} + T'_{\mu\nu}) &= (T_{\nu\mu} + T'_{\nu\mu}) \qquad \sum_\mu \frac{dT'_{\mu\nu}}{dx_\mu} = 0 \\
\int T'_{4\mu}\,d^3x &= 0
\end{aligned}\right\} \tag{11.28}$$

These stipulations are consistent with the degree of freedom in the definition of $\mathscr{L}$ which was discussed in the previous section.

It is possible to derive the constants of motion, in a more satisfying manner than here indicated, using a systematic process of deduction from Hamilton's principle.

Covariance

The desirability of a covariant formulation has already been emphasized, and the Lagrangian scheme as developed earlier in this chapter was deliberately designed to be so. This is not true of the Hamiltonian formulation which is based on the definition of canonical momenta given by:

$$\pi^{(r)} = \frac{\partial \mathscr{L}}{\partial \dot{\eta}^{(r)}} = \frac{1}{ic} \frac{\partial \mathscr{L}}{\partial \eta_{,\,4}{}^{(r)}} \qquad (11.22)$$

The quantities so defined are obviously not Lorentz invariant. This creates a difficulty of principle in connection with the transition to quantum theory. As has already been pointed out, in connection with continuous systems generally, it is usual to introduce quantum rules by specifying the values of the commutators of the operators representing the conjugate variables.* If the quantum behaviour is to be described in relativistic terms these commutators must be invariant; in terms of the above definition they will not be so.

A solution to the difficulty is to adopt a more general definition of $\pi^{(r)}$:

$$\pi^{(r)} = \sum_{\mu} n_{\mu} \frac{\partial \mathscr{L}}{\partial \eta_{,\,\mu}{}^{(r)}} \qquad (11.29)$$

where n_{μ} represents an arbitrary *time-like* direction in Minkowski 4-space.

i.e.,
$$\sum_{\mu} n_{\mu}{}^2 = -1 \qquad (11.30)$$

The restriction to a time-like direction is connected with the requirement to allow field disturbances to propagate only between events with a time-like separation.

* This is only true of fields associated with particles obeying Bose statistics (i.e., with integral spin). For those conforming to Fermi–Dirac statistics (half integral spin) the quantum behaviour is introduced by specifying the value of an *anti*-commutator.

This procedure can be developed to give a consistent covariant formalism, but here we shall content ourselves with having pointed out the difficulty involved and also the general direction of its resolution.

Summary

The brief sketch given in this chapter of the Lagrangian and Hamiltonian background to field theories can only be an outline guide to the subject. The intention is to emphasize the wide generality of the methods of analytical mechanics, which were developed in the first place as an alternative to Newton's laws in describing the behaviour of particles. The detailed working out of field theories is a long and complicated process, but their formulation is a comparatively simple and elegant matter. Naturally many difficulties have been glossed over in such a simplified account, but the essential structure should be apparent.

Examples

1. p_θ and θ are the angular momentum and the deflection of a simple pendulum. Plot the paths in phase space (i.e. p_θ versus θ) for various total energies including some in which θ exceeds $\pi/2$.

2. Newton's equations of motion for a particle in a two-dimensional conservative force field are

$$m\frac{d^2x}{dt^2} = -\frac{\partial V}{\partial x} \quad \text{and} \quad m\frac{d^2y}{dt^2} = -\frac{\partial V}{\partial y}$$

Derive from these the equations of motion in polar (r, θ) coordinates and compare with those derived directly from the Lagrangian equations of motion expressed in polar co-ordinates.

3. A particle of mass m is attracted to a centre by a force which is proportional to the distance from the centre. Show by use of Lagrange's equations of motion that the motion can always be resolved into three simple harmonic motions of the same period along three mutually perpendicular axes.

4. Obtain the Lagrangian and the equations of motion for a double pendulum with lengths l_1 and l_2 and masses m_1 and m_2. Show that the same equations can be obtained by a direct application of Newton's laws of motion.

5. Express Kepler's second law of motion for conservative central forces (i.e. $r^2\dot\theta = $ constant) in cartesian co-ordinates in a plane. Obtain the same expression by use of the Lagrangian expressed in cartesian co-ordinates.

6. A body falling freely 100 metres from rest at the equator is deflected by about 2 cm. Confirm this and state in which geographical direction the deflection takes place.

   Consider the possibility of using such an experiment to demonstrate the equivalence of gravitational and inertial mass.

7. It is said that the circulating winds due to an anticyclone pressure system are due to Corioli's forces. Sketch the wind system for

such a pressure system (a) in the northern hemisphere, (b) in the southern hemisphere. Take particular care about signs in the formulae and explain why the pressure system tends to be maintained. Why are anticyclones not observed near the equator?

8. Show that the components of the Lorentz force

$$\mathbf{F} = q\left(\mathbf{E} + \frac{1}{c}\mathbf{v} \wedge \mathbf{B}\right)$$

can be expressed in terms of the potentials ϕ and $\mathbf{A}$ in the form

$$F_i = \left(\frac{d}{dt}\frac{\partial}{\partial \dot{x}_i} - \frac{\partial}{\partial x_i}\right)q\left(\phi - \frac{1}{c}\mathbf{v}.\mathbf{A}\right)$$

Hence or otherwise confirm the validity of the usual Lorentz term in the Lagrangian.

9. Two bodies of reduced mass μ and relative separation $\mathbf{r}$ interact with a central force described by the potential $V(r)$. The total energy is E and the total angular momentum is $\mathbf{L}$. Show that the equation of the orbit is given by

$$d\theta = \frac{L}{\mu r^2}\left[\frac{2}{\mu}(E - V - L^2/2\mu r^2)\right]^{-\frac{1}{2}} dr$$

Hence, or otherwise, show that for an inverse square law of force and $E < 0$ that the orbit is an ellipse.

$$\left(\int dx/(ax^2 + bx + c)^{\frac{1}{2}} = \left((-a)^{\frac{1}{2}}\cos^{-1}\left[-\frac{(b + 2ax)}{(b^2 - 4ac)^{\frac{1}{2}}}\right]\right)\right)$$

10. Show that the Larmor precession frequency, ω_L, of the electrons in an atom in a magnetic field B_0 is given by $eB_0/2mc$ because $\omega_L \ll \omega_0$ where ω_0 is the orbital frequency of the electrons. Make a rough estimate of whether the Larmor theorem is valid for positronium (the atom consisting of a +ve and a −ve electron) in a field of 10,000 gauss.

11. If the kinetic and potential energies take the forms

$$T = \sum_i f_i(q_i)\dot{q_i}^2 \quad \text{and} \quad V = \sum_i V_i(q_i)$$

show that the Lagrangian equations of motion separate and are immediately integrable.

12. A particular molecule can be regarded as two masses m connected by a spring of constant k and unstretched length a. Show, using Lagrange's formulation, that if a molecule rotates with an angular momentum $\mathbf{L}$ and vibrates at the same time, the rotation is not uniform but is modulated at a frequency

$$\sqrt{\frac{2k}{m}}\left(1 + \frac{3L^2}{kma^4}\right)$$

(Assume that the amplitude of vibration is $\ll a$ and $\mathbf{L}$ is such that the centrifugal stretching is $\ll a$.) Discuss the spectroscopic consequences of such interdependence of rotational and vibrational motion.

13. A uniform bar of mass M and length $2l$ is suspended by one end from a spring of constant k. The spring can only stretch in a vertical direction and the bar can only swing freely in one vertical plane. Show that the same equations of motion are obtained by the Lagrangian and the Hamiltonian formulations.

14. Show that the shortest distance between two points in a plane is a straight line. Compare this result with that derived from the Lagrangian equations of motion for a free particle moving in a plane.

15. A particle of mass m falls under gravity from rest at $t = 0$ to a lower point at $t = T$. Consider the path $x = -\frac{1}{2}gt^2 + \alpha t(t - T)$ and show explicitly that Hamilton's principle demands that $\alpha = 0$, thus illustrating the correctness of the principle in this simple case.

16. A particle of mass m falls under gravity between points A and B separated by a vertical height h and horizontal distance d. Determine the path which would result in the shortest time of fall. (The brachistochrone problem.)

17. Show that the explicit form of the solution of the Hamilton–Jacobi equation for the problem of a linear harmonic oscillator of mass m, restoring force constant μ and displacement q is

$$S = \alpha\sqrt{\frac{m}{\mu}}\left[\sin^{-1}\frac{\mu}{2\alpha^2}q + \frac{\mu}{2\alpha^2}q\left(1 - \frac{\mu}{2\alpha^2}q^2\right)^{\frac{1}{2}}\right] - \alpha t$$

Identify the momentum in the transformed system from this value of S.

18. For the harmonic oscillator of question 17 assume the motion is given by

$$q = \sqrt{\frac{2E}{\mu}} \sin\left(\sqrt{\frac{\mu}{m}} t + \delta\right)$$

where E is the total energy. Hence evaluate S directly from the Lagrangian and thus identify α.

19. Apply the Hamilton–Jacobi method to the problem of question 9. (Hint: obtain H for plane polars, try $S = W(r) + \alpha_\theta\theta + \alpha t$, find $p_\theta' = \dfrac{\partial S}{\partial \alpha_\theta} = $ const., hence the required result.)

20. Prove that $\ddot{p}_i = \left[[p_i, H], H\right]$ if $\dfrac{\partial H}{\partial t} = 0$.

21. Show by direct evaluation that the Poisson bracket $[l_x, l^2] = 0$ where l_x is an angular momentum and l^2 is the total square angular momentum ($= l_x^2 + l_y^2 + l_z^2$).

22. It may be shown that for two *operators* $(AB - BA) = \alpha[A, B]$ where the expression on the right is the classical Poisson bracket and α is a constant. Show that this relation is satisfied for the operators x_i and p_i, provided p_i is interpreted as a constant $\times \dfrac{\partial}{\partial x_i}$.

23. Show that the equation of motion of a one-dimensional elastic solid, $\rho\ddot{\eta} = E\dfrac{d^2\eta}{dx^2}$, follows as a limiting case of a line of masses m linked by springs of constant k at intervals a. Show also that it follows from the Lagrangian density

$$\mathscr{L} = \tfrac{1}{2}\left\{\rho\dot{\eta}^2 - E\left(\frac{d\eta}{dx}\right)^2\right\}$$

24. Deduce the equations of motion corresponding to the Lagrangian density functions given on p. 112. Derive the corresponding Hamiltonian densities.

25. Verify that the correct electromagnetic field equations are obtained assuming a Lagrangian density of the form

$$\mathscr{L} = \alpha \sum_\mu \sum_\nu \left(\frac{dA_\mu}{dx_\nu}\right)^2$$

26. A gas at NTP consists of particles having the same mass as electrons but which behave mechanically as hard spheres of radius 10 Å. Consider whether or not the situation can be described adequately by classical non-relativistic mechanics.

Bibliography

This is a selected list of useful works. References to a number of accounts of more restricted topics are included as footnotes to the text.

(a) Works giving a critical examination of the bases of mechanics:

A. D'ABRO, *The Evolution of Scientific Thought*, Dover Press, New York, 1950

S. BANACH, *Mechanics*, Polish Mathematical Society, Warsaw, 1951

P. W. BRIDGMAN, *The Logic of Modern Physics*, Macmillan, New York, 1960

R. DUGAS, *Histoire de la Mécanique*, Editions du Griffon, Neuchâtel, 1950

R. B. LINDSAY and H. MARGENAU, *Foundations of Physics*, Dover Press, New York, 1957

E. MACH, *The Science of Mechanics*, Open Court, Illinois, 1942

J. L. SYNGE and B. A. GRIFFITH, *Principles of Mechanics*, McGraw-Hill, New York, 1959

W. YOURGRAU and S. MANDELSTAM, *Variational Principles in Dynamics and Quantum Theory*, Pitman, 1960

(b) Treatises covering approximately the same range of topics as the present volume, but at a more detailed or advanced level:

H. C. CORBEN and P. STEHLE, *Classical Mechanics*, John Wiley, New York, 1960

H. GOLDSTEIN, *Classical Mechanics*, Addison-Wesley, Camb., Mass., 1960

C. LANCZOS, *The Variational Principles of Mechanics*, University of Toronto Press, 1949

E. T. WHITTAKER, *Analytical Dynamics*, C.U.P., 1960

(c) More comprehensive works containing sections of interest:

L. BRILLOUIN, *Les Tenseurs en Mécanique et en Elasticité*, Masson, Paris, 1949

G. JOOS, *Theoretical Physics*, Blackie, 1951

R. B. LINDSAY, *Concepts and Methods of Theoretical Physics*, van Nostrand, New York, 1951

D. H. MENZEL, *Mathematical Physics*, Dover Press, New York, 1961

P. M. MORSE and H. FESHBACH, *Methods of Theoretical Physics* (Vol. I), McGraw-Hill, New York, 1953

A. SOMMERFELD, *Lectures on Theoretical Physics* (Vols I, II and III), Academic Press, New York, 1950–2

R. C. TOLMAN, *The Principles of Statistical Mechanics*, O.U.P., 1950

(d) Mathematical studies of the calculus of variations and its applications at a level suitable for further reading:

C. FOX, *An Introduction to the Calculus of Variations*, O.U.P., 1950

W. S. KIMBALL, *Calculus of Variations*, Butterworth, 1952

(e) Works devoted wholly or partly to special relativity theory:

J. AHARONI, *The Special Theory of Relativity*, O.U.P., 1959

P. G. BERGMANN, *An Introduction to the Theory of Relativity*, Prentice Hall, New York, 1942

H. DINGLE, *The Special Theory of Relativity*, Methuen, 1946

W. H. McCREA, *Relativity Physics*, Methuen, 1949

C. MØLLER, *The Theory of Relativity*, O.U.P., 1952

(f) Works devoted to classical field theory:

S. KAHANA and H. R. COISH, 'Classical Meson Theory', *Am. J. Phys.* **24**, 225–39 and 390–9 (1956)

L. LANDAU and E. LIFSCHITZ, *The Classical Theory of Fields*, Pergamon, Oxford, 1962

W. K. H. PANOFSKY and M. PHILLIPS, *Classical Electricity and Magnetism*, Addison-Wesley, Camb., Mass., 1955

(g) Treatises on quantum field theory with introductory accounts of the classical foundations:

E. M. CORSON, *Introduction to Tensors, Spinors and Relativistic Wave Equations*, Blackie, 1953

J. C. GUNN, *Theory of Radiation*, Rep. on Prog. in Physics, Vol. 18, Phys. Soc., London, 1955

J. M. JAUCH and F. ROHRLICH, *The Theory of Photons and Electrons*, Addison-Wesley, Camb., Mass., 1955

F. MANDL, *Introduction to Quantum Field Theory*, Interscience, New York, 1959

S. S. SCHWEBER, *An Introduction to Relativistic Quantum Field Theory*, Row, Peterson and Co., New York, 1961

G. WENTZEL, *Quantum Theory of Fields*, Interscience Publishers, New York, 1949

Index